T0145336

Studies in Computational Intelligence

Volume 920

The series "Studies in Computational Intelligence" (SCI) publishes new developments and advances in the various areas of computational intelligence—quickly and with a high quality. The intent is to cover the theory, applications, and design methods of computational intelligence, as embedded in the fields of engineering, computer science, physics and life sciences, as well as the methodologies behind them. The series contains monographs, lecture notes and edited volumes in computational intelligence spanning the areas of neural networks, connectionist systems, genetic algorithms, evolutionary computation, artificial intelligence, cellular automata, self-organizing systems, soft computing, fuzzy systems, and hybrid intelligent systems. Of particular value to both the contributors and the readership are the short publication timeframe and the world-wide distribution, which enable both wide and rapid dissemination of research output.

Indexed by SCOPUS, DBLP, WTI Frankfurt eG, zbMATH, SCImago.

All books published in the series are submitted for consideration in Web of Science.

More information about this series at http://www.springer.com/series/7092

Stefka Fidanova
Editor

Recent Advances in Computational Optimization

Results of the Workshop on Computational Optimization WCO 2019

Editor
Stefka Fidanova
Institute of Information and Communication
Technology
Bulgarian Academy of Sciences
Sofia, Bulgaria

ISSN 1860-949X ISSN 1860-9503 (electronic)
Studies in Computational Intelligence
ISBN 978-3-030-58886-1 ISBN 978-3-030-58884-7 (eBook)
https://doi.org/10.1007/978-3-030-58884-7

This Springer imprint is published by the registered company Springer Nature Switzerland AG
The registered company address is: Gewerbestrasse 11, 6330 Cham, Switzerland

Organization

Workshop on Computational Optimization (WCO 2019) is organized in the framework of Federated Conference on Computer Science and Information Systems FedCSIS-2019.

Conference Co-chairs

Stefka Fidanova, IICT-BAS (Bulgaria)
Antonio Mucherino, IRISA (Rennes, France)
Daniela Zaharie, West University of Timisoara (Romania)

Program Committee

Germano Abud, Universidade Federal de Uberlândia, Brazil
Tibérius Bonates, Universidade Federal do Ceará, Brazil
Mihaela Breaban, University of Iasi, Romania
Aritanan Gruber, Universidade Federal of ABC, Brazil
Khadija Hadj Salem, University of Tours—LIFAT Laboratory, France
Hiroshi Hosobe, National Institute of Informatics, Japan
Andrzej Janusz, University of Warsaw, Poland
Carlile Lavor, IMECC-UNICAMP, Campinas, Brazil
Mircea Marin, West University of Timisoara, Romania
Flavia Micota, West University Timisoara, Romania
Ionel Muscalagiu, Politehnica University Timisoara, Romania
Camelia Pintea, Tehnical University Cluj-Napoca, Romania
Stefan Stefanov, Neofit Rilski University, Bulgaria
Catalin Stoean, University of Craiova, Romania
Antanas Zilinskas, Vilnius University, Lithuania

Preface

Many real-world problems arising in engineering, economics, medicine and other domains can be formulated as optimization tasks. Every day, we solve optimization problems. Optimization occurs in the minimizing time and cost or the maximization of the profit, quality and efficiency. Such problems are frequently characterized by non-convex, non-differentiable, discontinuous, noisy or dynamic objective functions and constraints which ask for adequate computational methods.

This volume is a result of very vivid and fruitful discussions held during the Workshop on Computational Optimization. The participants have agreed that the relevance of the conference topic and quality of the contributions have clearly suggested that a more comprehensive collection of extended contributions devoted to the area would be very welcome and would certainly contribute to a wider exposure and proliferation of the field and ideas.

The volume includes important real problems like modeling of physical processes, wildfire and flood risk modeling, workforce planning, parameter settings for controlling different processes, optimal electrical vehicle modeling, bioreactor modeling, design of VLSI. Some of them can be solved applying traditional numerical methods, but others need huge amount of computational resources. Therefore, for them are more appropriate to develop an algorithms based on some metaheuristic method like evolutionary computation, ant colony optimization, particle swarm optimization, bee colony optimization, constrain programming, etc.

Sofia, Bulgaria
May 2020

Stefka Fidanova
Co-Chair, WCO'2019

Contents

Validation and Optimization of Dam Break Flood Risk Mapping Based on Field Test Cases in Armenia 1
Nina Dobrinkova, Alexander Arakelyan, Evangelos Katsaros, and Sean Reynolds

Fire Simulator Capable to Analyze Fire Spread in Real Time with Limited Field Weather Data. Case Study—Kresna Fire (2017) 33
Nina Dobrinkova and Adrián Cardil

Utilizing Minimum Set-Cover Structures with Several Constraints for Knowledge Discovery on Large Literature Databases 49
Jens Dörpinghaus, Carsten Düing, and Vera Weil

Evaluation of Optimal Charging Station Location for Electric Vehicles: An Italian Case-Study 71
Edoardo Fadda, Daniele Manerba, Gianpiero Cabodi, Paolo Camurati, and Roberto Tadei

InterCriteria Analysis of the Evaporation Parameter Influence on Ant Colony Optimization Algorithm: A Workforce Planning Problem 89
Olympia Roeva, Stefka Fidanova, and Maria Ganzha

Caterpillar Alignment Distance for Rooted Labeled Caterpillars: Distance Based on Alignments Required to Be Caterpillars 111
Yoshiyuki Ukita, Takuya Yoshino, and Kouichi Hirata

ICrA Over Ordered Pairs Applied to ABC Optimization Results 135
Olympia Roeva and Dafina Zoteva

A Game Theoretical Approach for VLSI Physical Design Placement 149
Michael Rapoport and Tami Tamir

Application of Information Systems and Technologies in Transport 173
Kristina Pavlova and Vladimir Ivanov

Online Algorithms for 1-Space Bounded Cube Packing and 2-Space Bounded Hypercube Packing 183
Łukasz Zielonka

Author Index 199

Validation and Optimization of Dam Break Flood Risk Mapping Based on Field Test Cases in Armenia

Nina Dobrinkova, Alexander Arakelyan, Evangelos Katsaros, and Sean Reynolds

Abstract The Alliance for Disaster Risk Reduction (ALTER) project began in February of 2018 with the goal of establishing public-private partnerships in Armenia to address flood risks that stem from water and mine dam failures. During the project duration, targeted work packages including extensive research, consensus building, technological implementation, and dissemination have been implemented covering flooding risk analyses and modeling for tailing storage facility and reservoir dam failures in the pilot areas of the Syunik and Lori regions. Based on the risk identification for the Kapan, Sisian, and Akhtala communities, disaster risk management and response plans were developed, tested, and refined during table top exercises conducted in those communities. Disaster response field exercises of unprecedented scale were implemented in Sisian, Kapan, and Akhtala involving more than 1300 participants. Public-private partnership MoUs were the logical outcome of the disaster risk management efforts at the local level.

Keywords Dam failure · ALTER project · Earthquake · Flood risk maps

---

N. Dobrinkova (✉)
IICT-BAS, Acad. Georgi Bonchev str. Bl.2, 1113 Sofia, Bulgaria
e-mail: nido@math.bas.bg; ninabox2002@gmail.com

A. Arakelyan · S. Reynolds
American University in Armenia, 40 Marshal Baghramyan Ave, 0019 Yerevan, Armenia
e-mail: alexander.arakelyan@aua.am; alex.arakelyan@outlook.com

S. Reynolds
e-mail: sreynolds@aua.am

A. Arakelyan
Institute of Geological Sciences, NAS, Armenia, 24a Marshal Baghramyan Ave, 0019 Yerevan, Armenia

E. Katsaros
European University Cyprus, Diogenis Str. 6, 2404 Nicosia, CY, Cyprus
e-mail: e.katsaros@research.euc.ac.cy

S. Fidanova (ed.), *Recent Advances in Computational Optimization*, Studies in Computational Intelligence 920,
https://doi.org/10.1007/978-3-030-58884-7_1

1 Introduction

The Alliance for Disaster Risk Reduction project (ALTER project) has been designed in the framework of DG ECHO external line call. These types of projects have as the main goal to address cooperation between the EU and third party countries. Main idea of the project was best practices transfer from the EU to an external neighboring country. For the ALTER project, that selected country was Armenia.

The project main objectives were to create a comprehensive research about the dam status in selected zones in Armenia by calculating the risk of flood events in cases of dam failure. The public private partnerships establishment was the other main goal of the project in order to secure the future sustainability of the project outcomes. The risks evaluated for the test areas were designed in a way to increase the resilience in areas of Armenia that face the risk from floods originating after earthquakes. Methods, tools, know-how and experience from Greece, Bulgaria and Cyprus have been shared with Armenian partners. The partnership of the Armenian government and local stakeholders gave an opportunity for the consortia to work on larger scale at the selected test areas. The project focused on three pilot areas in Armenia where dams and other activities such as mining processes were in place and the risks to local communities was evaluated as high. The test areas selected were: Akhtala and Teghut areas of Lori Marz along the Shamlugh river; the Vorotan Cascade and its associated dams in the Syunik region; and the Voghji river basin of Syunik region.

In the paper, we will provide chronological information about the research works, field tests, and exercises used for optimization of local authority's reactions in cases of dam break flood events. The software developed for the visualization needs of the project has been implemented in only one of the test areas—Kapan with the idea to be further developed by the Armenian end users after all gaps are revealed during the testing and validation procedures with the table top and field exercises.

2 Study Area

One of the activities of the project ALTER was to identify the most suitable best practices on risks related to dams in earthquake zones available within and outside the consortia members. Full scale dam break flood risk mapping research, field exercises and software development were done for the Kapan and Voghji River Basin study area. This area is located about 300 km southeast of Yerevan and has a population of about 45,000. It contains some of Armenia's most intensive mining activities and two of Armenia's largest tailing dams—Artsvanik and Geghanoush. Additionally, the Geghi Reservoir upstream of Kapan was also included. The villages Kavchut, Andiokavan, Hamletavan, Shgharjik, Syunik and the Kapan Town are located in the immediate floodplain of the Geghi and Voghji Rivers. The village of Verin Giratagh and Nerkin Giratagh are not in the floodplain, however the only road access to these villages is through the floodplain below the Geghi dam. The two tailing dams also pose risk to Kapan's airport, which would be needed in case of an emergency, and the main highway connecting Armenia and Iran.

2.1 Geghi Reservoir

The Geghi reservoir is located in Syunik, the southernmost province of Armenia (Fig. 1). The reservoir is situated on the Geghi river, the left-bank tributary of the river Voghji. The maximum water level discharge occurs during the spring. Due to the high altitude nature of the area, snowmelt increases gradually as does the level of the river and the reservoir. Snowmelt typically occurs from March to August (Armenian State Hydrometeorological and Monitoring Service).

The surface of the Geghi reservoir is 50 ha and the elevation above sea level is nearly 1400 m. The height of the dam is 70 m and the length along the crest is 270 m. The total volume of reservoir is 15 million m^3, but the effective volume is about 12 million m^3 [1]. Nearly 4300 people would be affected by a dam break affecting the reservoir [2].

Fig. 1 The location of Geghi reservoir. The inset shows its location within Armenia. Background image: Sentinel-2, RGB composite

2.2 *Geghanoush Tailing Storage Facility (TSF)*

Geghanoush TSF is located in the gorge of the Geghanoush River, in the southern part of Kapan (Fig. 2). The difference of relative heights between the tailing dam, on one hand, and city buildings and transport infrastructure, on the other hand, is 75 m. In case the reservoir dam is broken due to an earthquake, the sliding mass could cover industrial and residential buildings, and as a result of barrage, the polluted water could flood central quarters of the city.

The existing Geghanoush Tailings Repository was designed in early 1960s and had been operated between 1962 and 1983, when the Kajaran Tailings Repository at Artsvanik was commissioned. The Geghanoush tailings repository was re-commissioned in 2006 after the completion of the diversion works and continues to be used today along with an upstream extension currently under construction. The volume of the tailing is 5.4 million m^3 and the dam height is 21.5 m [1].

Tailing and water dams in the appointed pilot area are hazardous hydro-technical structures because of their location in earthquake prone zone. In addition, a dam break could occur due to the technical condition of the dams and improper exploitation. Catastrophic floods are possible in the area as a result of dam failure. Therefore, the assessment of dam break consequences has a crucial meaning for emergency management and development for measures and action plans for stakeholders and respective authorities in Armenia.

Fig. 2 The location of Geghanoush tailing dam location in Armenia. Background image: Sentinel-2, RGB composite. Panoramic view of the dam (photo credit: «Haekshin» Company)

3 Methodology and Data Used for Simulations

3.1 Methodology

Despite their many beneficial uses and value, dams also considered as a hazardous infrastructure due to their possible failure and massive flooding in consequence. To mitigate these risks, potential dam failure event need to be modeled by analyzing dam breach properties and possible extent of the flooding formed in a result of dam break.

The two primary tasks in the analysis of a potential dam failure are the prediction of the reservoir outflow hydrograph and the routing of that hydrograph through the downstream valley to determine dam failure consequences. When populations at risk are located close to a dam, it is important to accurately predict the breach outflow hydrograph and its timing relative to events in the failure process that could trigger the start of evacuation efforts [3].

The comprehensive overview on dam breach characteristics analysis and peak outflow calculation equations can be found in details in the document entitled "Using HEC-RAS for Dam Break Studies" compiled by Hydrologic Engineering Center of US Army Corps in 2014 [4].

As for the tailing dams, these equations and models can't be directly applied. There are several scientific papers on the methods for the calculation of TSF failure outflow characteristics and maximum distance travelled by tailings. One of these methods developed by Concha Larrauri and Lall [5] who combined the lists of TSF failures compiled by Rico et al. [6] and Chambers and Bowker (TSF Failures 1915–2017 as of 16 August 2017) [7] and compared the results of the original linear regressions done by Rico et al. [6] with the results using the updated dataset.

The regression equations on dam breach and maximum outflow characteristics both for water reservoirs and TSFs are generally developed based on the statistical analysis of above-mentioned parameters in past dam break events. In this paper, TR-60 and TR-66 simplified dam breach outflow and routing models developed by the Engineering division of Soil Conservation Service of US Department of Agriculture [8] is used for Geghi dam break characteristics analysis, and method developed by Concha Larrauri and Lall [5] is used for Geghanoush TSF dam failure analysis.

Flood modeling basics refer to 1D and 2D models, which provide steady and unsteady flows, including the necessity of Manning N values usage.

There are many event types and phenomena that can lead to dam failure:

- Flood event
- Landslide
- Earthquake
- Foundation failure
- Structural failure
- Piping/seepage (internal and underneath the dam)
- Rapid drawdown of pool

- Planned removal
- Terrorism act.

Given the different mechanisms that cause dam failures, there can be several possible ways dam may fail for a given driving force/mechanism. In 1985 and in 2002 data has been analyzed a list of dam types [9, 10] versus possible modes of failure.

The reports from 1985 noticed that of all dam failures—34% were caused by overtopping, 30% due to foundation defects, 28% from piping and seepage, and 8% from other modes of failure. In the same report of dam failures are included earth/embankment dams, for which 35% have failed due to overtopping, 38% from piping and seepage, 21% from foundation defects, and 6% from other failure modes.

In our work we are doing analysis of a potential dam failure. The prediction of the reservoir outflow hydrograph and the routing of that hydrograph through the downstream valley are evaluated to determine dam failure consequences. There are calculated results about the risk of the population located close to the dam, it is important to accurately predict the breach outflow hydrograph and its timing relative to events in the failure process that could trigger the start of evacuation efforts [6].

3.2 *Data Sets Used for the Modelling*

3.2.1 Hydro-Meteorological Observation Data

Flood formation and its behavior is highly depends from hydro-meteorological conditions of the territory. Rainfall intensity and duration, snowmelt, air temperature and other meteorological factors are key drivers in flood development process. Hydro-meteorological monitoring within the territory of Armenia is conducted by Hydromet Service of the Ministry of Emergency Situations of Armenia (Fig. 3).

There are 2 operational meteorological stations within Voghji River Basin: Kajaran and Kapan (Table 1).

Thermal conditions normally decrease in the Voghji Basin as altitude increases. Multiyear annual average air temperature is in Kajaran is 6.8 °C and in Kapan is 12.3 °C (Table 2).

Rainfall generally increases by altitude in the basin (Table 3).

The average annual relative humidity is 50–60%, and less than 30% at low altitudes (up to 1000 m). Frost-free days vary by altitude—annually from 260 (at the altitude of 700 m) to 50 days (higher than 3000 m). The annual average relative humidity is 60–80% (over 2600 m), and at lower altitudes—up to 30% (up to 1000 m).

Permanent snow cover starts at altitudes of 1200 m and it lasts for 35–165 days. The snow depth is 15–180 cm. It lasts 1–1.5 months at altitudes of up to 1500 m, and 6.5–7 months at altitudes of 3000 m and higher. The depth of snow cover is 15–20 cm at altitudes of 1300–1500 m and 120–180 cm at altitudes of 3000 m and higher (from

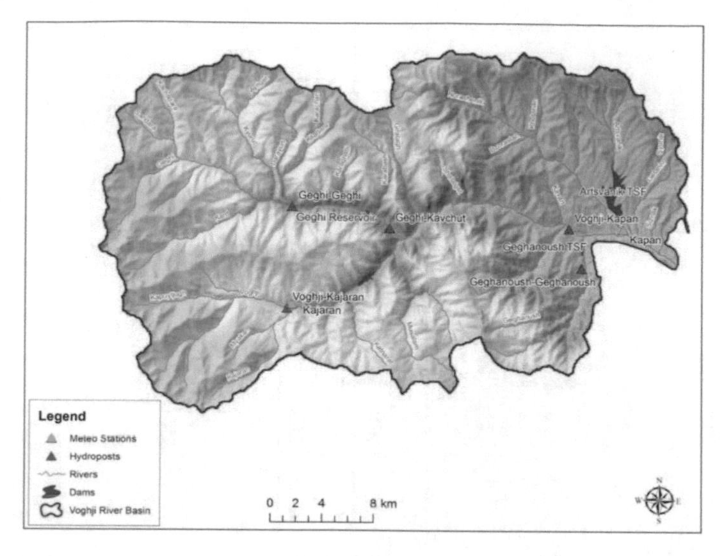

Fig. 3 Hydro-meteorological monitoring posts within Voghji river basin

Table 1 Operational monitoring stations within Voghji river basin

No	Name of station	Latitude	Longitude	H (m)	Observation period
1	Kajaran	39° 09′ 10″	46° 09′ 33″	1843	1975–present
2	Kapan	39° 12′ 15″	46° 27′ 44″	705	1936–present

place to place a 300 cm thick snow cover is formed, due to winds occurring in concavities). Evaporation drops to 482–220 mm as altitude increases in the Voghji River Basin. The highest value of evaporation, 500–480 mm, is observed at altitudes up to 800 m. There are 3 operational hydrological monitoring posts within Voghji River Basin: Voghji-Kajaran, Voghji-Kapan and Geghi-Kavchut. Data of closed monitoring posts of Geghi-Geghi and Geghanoush-Geghanoush were analyzed as well due to their importance for the Geghi reservoir and Geghanoush tailings dam break modeling (Tables 4 and 5).

3.2.2 Elevation Data and Its Derivatives

Elevation data has a crucial meaning in each flood modeling process. There are various free digital elevation models (DEMs) available online (SRTM, ASTER, ALOS), the spatial resolution of which is ~30 m. This resolution is not enough for detailed flood mapping in mountainous areas.

Georisk CJSC provided linear shapefile of elevation isolines of 1:10,000 scale. From this shapefile, 5 m resolution DEM of studied area was calculated using Topo to Raster interpolation tool of ArcGIS Spatial Analyst toolbox (Figs. 4 and 5).

Table 2 Annual and monthly average air temperatures in the Voghji river basin, °C

Meteorological station	Absolute altitude (m)	Month												Year
		I	II	III	IV	V	VI	VII	VIII	IX	X	XI	XII	
Kajaran	1980	−3.4	−3.0	0.5	5.7	10.2	14.2	17.1	16.6	13.3	8.2	3.2	−1.0	6.8
Kapan	704	0.8	2.4	6.3	12.3	16.1	20.4	23.7	23.1	19.0	13.0	7.5	2.9	12.3

Table 3 Intra-annual distribution of atmospheric precipitation in the Voghji river basin, mm

Meteorological station	Absolute altitude (m)	Month												Year
		I	II	III	IV	V	VI	VII	VIII	IX	X	XI	XII	
Kajaran	1980	44	51	74	84	85	49	23	21	31	52	49	41	605
Kapan	704	25	31	59	75	94	66	31	28	41	49	40	25	565

Table 4 Hydrological monitoring posts within Voghji river basin

No	Water object name	Name of station	Coordinates	
			Latitude	Longitude
1	Voghji river	Kajaran	39° 08′ 59″	46° 09′ 16″
2	Voghji river	Kapan	39° 12′ 18″	46° 24′ 43″
3	Geghi river	Kavchut	39° 12′ 23″	46° 14′ 50″
4	Geghi river	Geghi	39° 13′ 21″	46° 9′ 36″
5	Geghanoush river	Geghanoush	39° 10′ 35″	46° 25′ 24″

Geomorphometric parameters (slope, aspect and shaded relief) were derived from DEM (Fig. 7).

Using ArcHydro Tools, following raster layers were calculated from DEM:

- Filled DEM (hydrologically-corrected);
- Flow Direction;
- Flow Accumulation;
- Streams (defined and segmented);
- Catchments GRID.

From these layers, catchment polygon and drainage line vector layers were obtained (Fig. 6).

3.2.3 Land-Cover and Land-Use

A detailed land-use and land-cover (LULC) maps is an important component for flood modeling in order to understand the extent of the flood and vulnerability from it, decision-making, as well as a source of baseline information for environmental research. The American University of Armenia GIS and Remote Sensing Lab proposed to carry out a LULC mapping exercise for the Voghji River basin based on freely available data from the novel Sentinel-1 and Sentinel-2 missions operated jointly by the European Space Agency (ESA) and the European Commission [11].

For open water (including Geghi reservoir) and tailing ponds Sentinel-1 (SAR) data was used. After obtaining the results of open water and tailing ponds then they were superimposed on the other classes. The final map is shown in Fig. 7.

3.2.4 Other Spatial Data

GIS layers available in AUA and Institute of Geological Sciences geodatabases (infrastructure, buildings, administrative units, water objects, monitoring sites location, etc.), as well as CAD drawings of Geghi and Geghanoush Dam areas provided by National University of Architecture and Construction and Georisk CJSC are used in dam breach analysis, flood hazard assessment and mapping.

Table 5 Flow characteristics in the hydrological monitoring posts within Geghi river basin

River-post	Discharge (m^3/s)													
	I	II	III	IV	V	VI	VII	VIII	IX	X	XI	XII	Annual average	Maximum
Geghi-Geghi	1.5	1.5	1.9	5.5	12.7	12.9	6.9	3.3	2.3	2.0	1.8	1.6	4.5	37.7
Geghi-Kavchut	1.4	1.5	2.3	5.9	12.5	12.0	6.3	3.0	2.1	1.9	1.7	1.5	4.4	87.5
Geghanoush-Geghanoush	0.2	0.2	0.7	1.9	1.8	0.9	0.4	0.3	0.3	0.3	0.3	0.2	0.6	21.3
Voghji-Kajaran	0.5	0.5	0.8	3.0	7.7	11.5	7.2	2.5	1.0	0.7	0.6	0.5	3.0	43.9
Voghji-Kapan	2.4	2.6	4.6	14.5	28.7	28.5	15.6	6.4	3.9	3.6	3.2	2.7	9.7	270.0

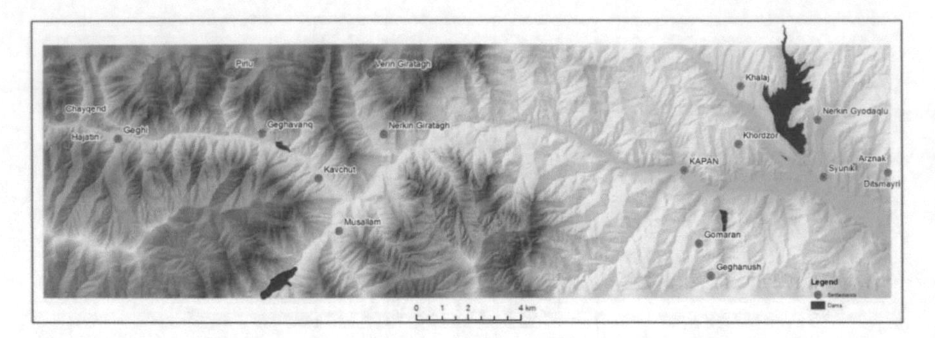

Fig. 4 5 m DEM of studied area

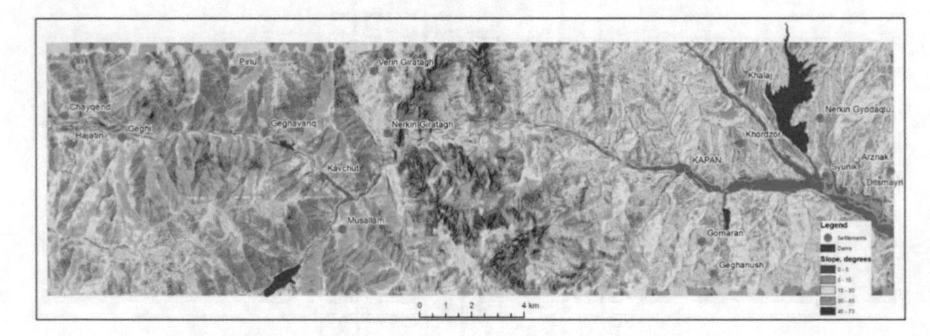

Fig. 5 Slope map of studied area

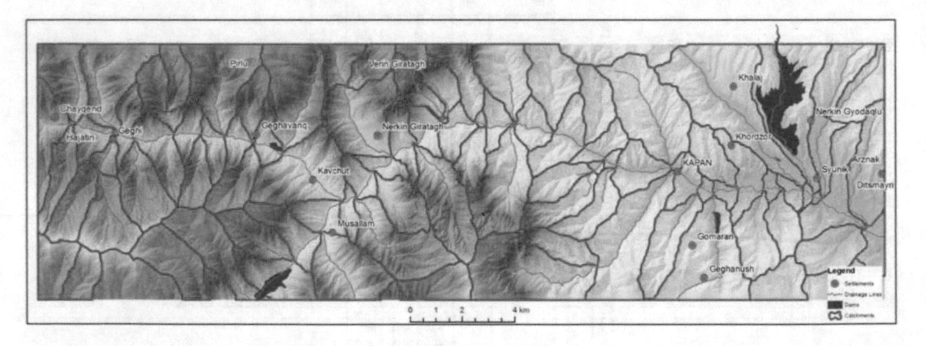

Fig. 6 Catchments and drainage network delineated using ArcHydro

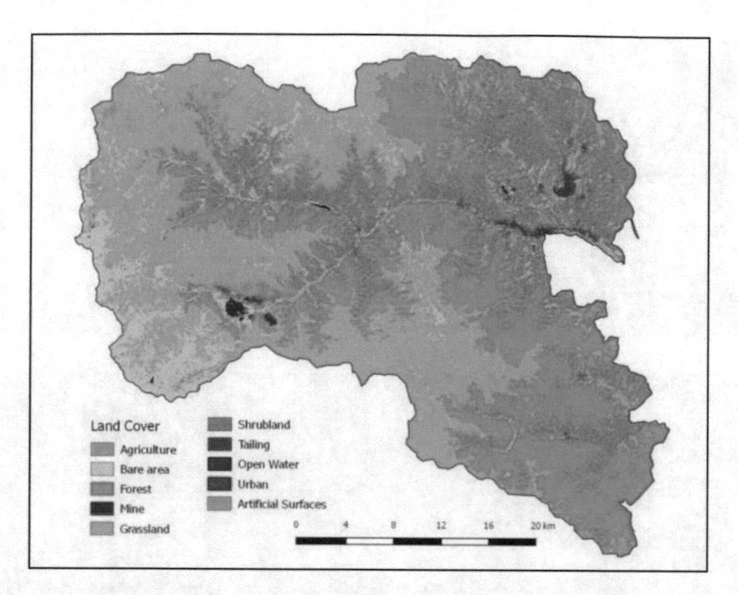

Fig. 7 Land-cover and land-use map of Voghji river basin

4 Simulation Results

4.1 Input Data

The analysis of dam-break flood characteristics was performed using the Volna 2.0 and flood mapping was conducted through the Spatial Analyst and 3D Analyst extensions of ArcGIS 10.x. Data from various sources including historic river and flood data, recent land use/land cover data, and updated cadaster data was used for analysis and mapping.

The results are presented in form of tables, maps and GIS layers.

The following information and datasets were used in dam-break flood modelling for Vorotan Cascade and Geghi Reservoirs:

- Hydro-technical characteristics of reservoirs and their dams (Fig. 8);
- Hydrographic and hydrologic characteristics of Voghji, Geghi and Vorotan Rivers;
- Digital elevation model of the studied area (with 5 m resolution);
- Satellite imagery (ESRI Base Map);
- Vector layers on geographic and land use characteristics of downstream area (Fig. 9).

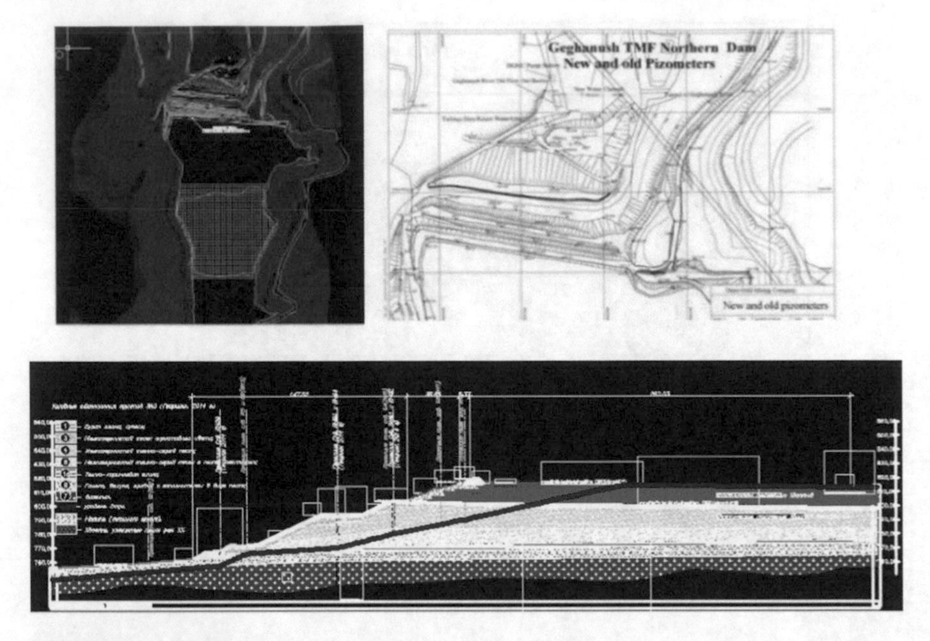

Fig. 8 Hydro-technical characteristics of reservoirs and dams

Fig. 9 Land cover map of studied area

4.2 *Software*

Volna 2.0 (software) and the Spatial Analyst and 3D Analyst extensions of ArcGIS 10.x were used for dam breach analysis, flood extent prediction, and mapping.

4.3 Method Used

Contour lines digitized from 1:10,000 topo maps of the study area were imported to ArcGIS and projected to WGS_1984_UTM_Zone_38N coordinate system. A digital elevation model (DEM) and triangular irregular network (TIN) were generated through interpolation of contour lines using Topo to Raster and Create TIN tools (Fig. 10).

Using the 3D Analyst extension, cross-sections of river valley were constructed based on the DEM (Fig. 11).

Hydro-technical characteristics of the reservoirs and their dams as well as hydrological and geomorphometric characteristics of river cross-sections were imported

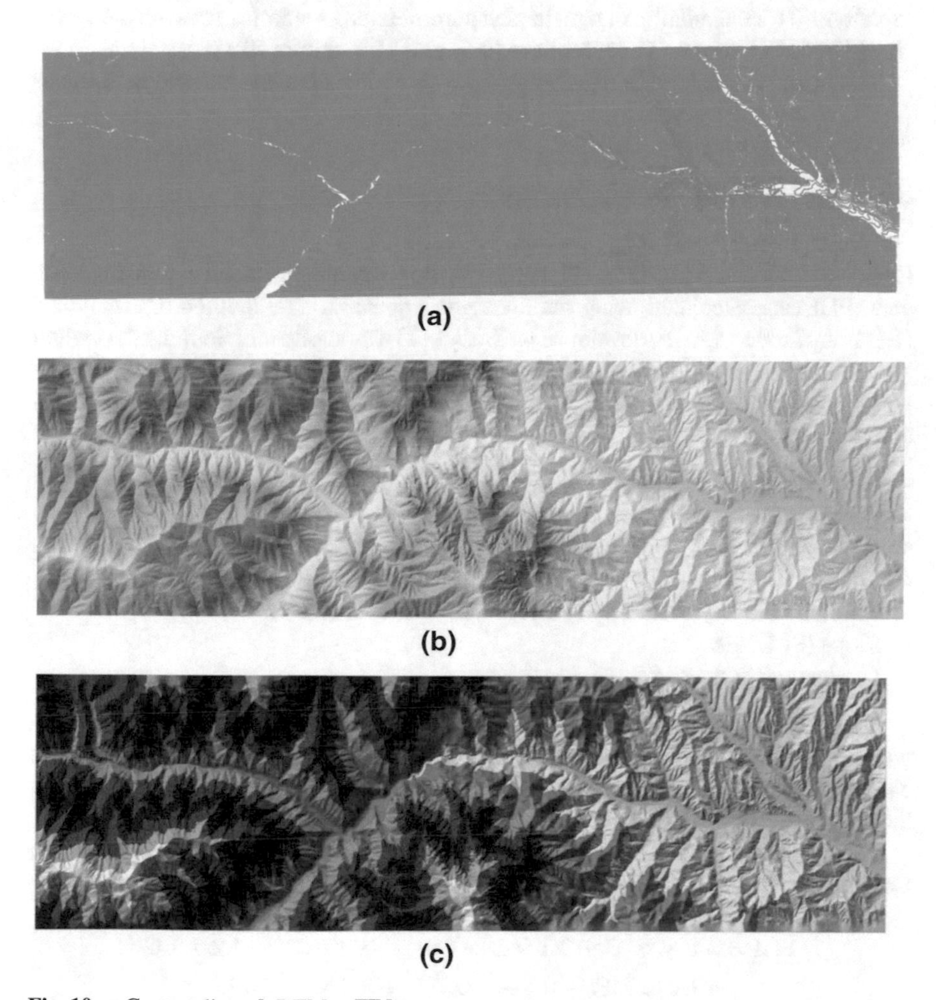

Fig. 10 **a** Contour lines, **b** DEM, **c** TIN

Fig. 11 Cross-sections used for dam break flood modelling

to Volna 2.0 for calculation of dam breach parameters. 3 scenarios for each dam were considered: full failure (1), half failure (0.5) and 10% failure (0.1).

The input parameters for dam break analysis Volna 2.0 are presented in Table 6.

4.4 Result 1. Flood Hazard Index

Flood hazard index (FHI) (Fig. 12) represents the probability of flood events in studied area. FHI was calculated using multi-criteria approach. The method of Analytical Hierarchy Process (AHP) developed by Saaty [12] was applied for defining the weight of each parameter.

The following parameters are accepted for flood hazard index calculation by above-mentioned method:

- Flow accumulation (FAC);
- Distance from drainage network (DIST);
- Elevation (ELEV);
- Land cover (LC);
- Rainfall intensity described by Modified Fourier Index (MFI);
- Slope (SLOPE);
- Geology (GEOLOGY).

The raster layers of above-mentioned parameters were developed, and respective weights were assigned to these parameters. Each of parameters were classified, and each class assigned with a rating from 2 to 10 (2 is less hazardous, 10 is most hazardous areas).

A flood hazard index was calculated through the following equation in Raster Calculator (ArcGIS) using the above-presented layers on flood hazard parameters:

$$\begin{aligned} \mathrm{FHI} = {} & 3.0 \times \mathrm{FAC} + 2.1 \times \mathrm{DIST} + 2.1 \times \mathrm{ELEV} + 1.2 \times \mathrm{LC} \\ & + 1.0 \times \mathrm{MFI} + 0.5 \times \mathrm{SLOPE} + 0.3 \times \mathrm{GEOLOGY} \end{aligned}$$

Table 6 Input data for dam break analysis in Volna 2.0 software (example for Geghi reservoir)

Dam cross-section characteristics		Unit	0-cross section							
Reservoir volume at the normal level	Wr	mln m^3	15							
Reservoir depth at the normal level	Hr	m	70							
Reservoir surface area at the normal level	Sr	mln m^2	0.5							
Reservoir width near the dam at the normal level	Br	m	270							
River depth at the lower reach of the hydrojunction	Hlr	m	1							
River width at the lower reach of the hydrojunction	Blr	m	12							
Flow velocity at the lower reach of the hydrojunction	Vlr	m/s	2							
Reservoir depth near the dam at the moment of dam break	Hdb	m	67							
Dam break level	Edb		1							
Dam breach height	h	m	0							
Water surface of absolute height at the normal level	Zr	m	1402							
Cross-sections on the river	N		8							
River cross-section characteristics		Unit	1 cs	2 cs	3 cs	4 cs	5 cs	6 cs	7 cs	8 cs
Distance of the cross-section from the dam	Lci	km	3	7	12	17	22	24	26	30
Normal flow										
Absolute height of water surface	Zbi	m	1258	1156.5	892	799	736	692	660	643
Depth	Hbi	m	0.5	1	1	1	1	1	1	1
Width	Bbi	m	30	11.5	11	14	11	15	21	21

(continued)

Table 6 (continued)

Dam cross-section characteristics		Unit	0-cross section							
Flow velocity	Vbi	m/s	2	2	2	2	2	2	2	2
Left bank										
First contour line value	z1	m	1260	1160	895	800	740	695	665	645
Distance from river centerline to the first contour line	B1	m	44	32	25	17	240	122	82	212
Second contour line value	z2	m	1265	1165	900	805	745	700	670	650
Distance from river centerline to the second contour line	B2	m	112	38	88	28	280	263	128	288
Third contour line value	z3	m	1270	1170	905	810	750	930	675	655
Distance from river centerline to the third contour line	B3	m	143	54	94	36	296	329	356	308

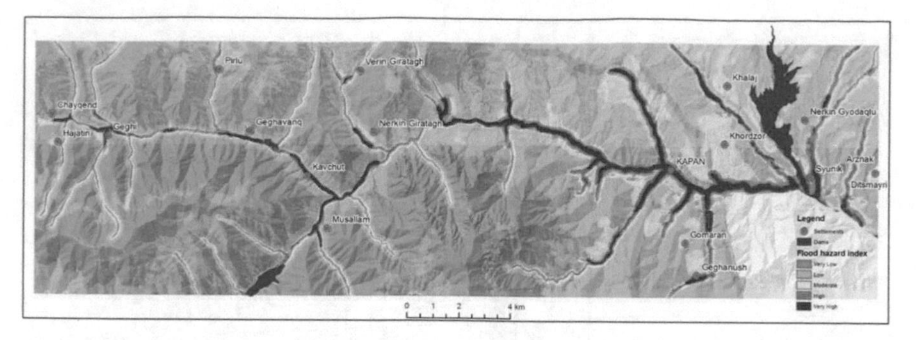

Fig. 12 Flood hazard index (FHI) map of studied area

4.5 Result 2. River Maximum Flow Analysis

Maximum flow probabilities were calculated using annual absolute maximum time-series for hydroposts of Voghji River Basin, mathematical-statistical methods applied in hydrology since 1983, StokStat 1.2 hydrologic software and Rybkin-Alekseev equations.

Table 7 presents calculated values of maximum flow of different probabilities for 5 hydrologic monitoring posts in Voghji River Basin: Geghi-Geghi, Geghi-Kavchut, Voghji-Kajaran, Voghji-Kapan and Geghanoush-Geghanoush.

On May 15, 2010 in a result of the accident in Geghi Reservoir, water level in Geghi and Voghji Rivers increased rapidly, and water flooded a large area downstream, including arable land and buildings in Kapan Town. Maximum discharge in Geghi-Kavchut hydropost was 87.5 m^3/s (which is greater than 0.01% probability flow by 1.5 times) and 133 m^3/s in Voghji-Kapan hydropost (which is greater than 1% probability flow).

In August 29, 1956, a mudflow with 270 m^3/s discharge was recorded. This discharge is 1.3 times greater than 0.01% probability maximum flow (Fig. 13).

4.6 Result 3. Dam Breach Maximum Outflow and Breach Hydrograph Calculation

The necessary input data for maximum outflow Q_{max} and hydrograph calculations using TR-60 and TR-66 models are listed in Table 8.

Theoretical breach width is calculated using following equation:

$$T = \frac{65\left(H_w{}^{0.35}\right)}{0.416}$$

Maximum outflow is calculated using following equation:

Table 7 Maximum flows of different probabilities calculated for hydroposts of Voghji river basin[a]

River-post	Q_0	C_v	C_s	Probability									
				0.01	0.1	1	2	5	10	25	50	95	99
Geghi-Geghi	23.85	0.336	0.159	56.31	50.38	43.40	40.92	37.31	34.27	29.14	23.69	11.03	6.06
Geghi-Kavchut	23.2	0.341	0.296	57.82	51.02	43.27	40.66	36.79	33.55	28.25	22.79	10.93	6.58
Voghji-Kajaran	17.5	0.373	1.38	62.31	50.70	38.82	35.17	30.21	26.23	20.69	16.06	9.86	8.88
Voghji-Kapan	52.8	0.469	1.05	203.12	166.72	128.58	116.20	99.36	85.99	66.17	48.59	20.61	14.17
Geghanoush-Geghanoush	8.5	0.62	1.359	44.02	34.96	25.58	22.68	18.73	15.56	11.14	7.34	2.28	1.39

[a] Q_0 is time-series average value; C_v is variation coefficient; C_s is asymmetry or skewness coefficient

Fig. 13 Flood event caused by an accident in Geghi reservoir (May 15, 2010), photo credit: https://www.a1plus.am

Table 8 Necessary input data for TR-60 and TR-66 models

Elevations		
Top of dam	4609.6	Ft* msl
Water surface@breach	4599.8	Ft msl
Average valley floor	4379.9	Ft msl
Wave berm	4489.9	Ft msl
Stability berm	4489.9	Ft msl
Length of dam@breach elev	886	Ft
Storage volume@breach elev	12,161	Ac Ft
Top width	32.8	Ft
Upstream slope above berm	2.5	:1
Upstream slope below berm	2	:1
Downstream slope above berm	2.5	:1
Downstream slope below berm	2	:1
US wave berm width	50	Ft
DS stability berm width	50	Ft

$$Q_{max} = 65\left({H_w}^{1.85}\right)$$

where H_w is height of the breach (ft).

Calculations for Geghi were made in US (Imperial) units and then converted to SI (Metric) units. Three scenarios of dam break are considered for Geghi Dam. The results are presented in Table 9.

The hydrographs for three dam-break scenarios are presented in Fig. 14.

Geghanoush TSF dam failure maximum outflow and tailing travel distance were calculated using the methodology developed by Concha Larrauri and Lall [5]:

$$V_F = 0.332 \times {V_T}^{0.95} \; D_{max} = 0.332 \times {H_F}^{0.545} \; H_f = H \times (V_F/V_T) \times V_F$$

Table 9 Calculated values of Geghi dam breach outflow and timesteps using TR-60 and TR-66 models

Timesteps (min)	Breach outflow (m^3/s)		
	Worst case	Average case	Best case
0	56.31[a]	56.31[a]	56.31[a]
6	39,700.6	13,895	3970
12	5968.7	12,936.2	7576.6
18	2339.6	10,081.2	8771.1
24	938.1	6099.2	8084.2
30	396.8	2977.4	6721.1
36	187.8	1478.1	5368.7
42	107.1	881.3	3680.2
48	75.9	592.0	2219.8
54	63.9	270.3	1493.1
60	59.2	239.9	773.8
66	57.4	186.5	593.4
72	56.7	160.0	378.3
78	56.5	133.9	235.1
84	56.4	108.0	163.6
90	56.3	82.1	92.1

[a]Maximum flow of 0.01% probability in Geghi-Geghi monitoring post (calculated in previous chapter)

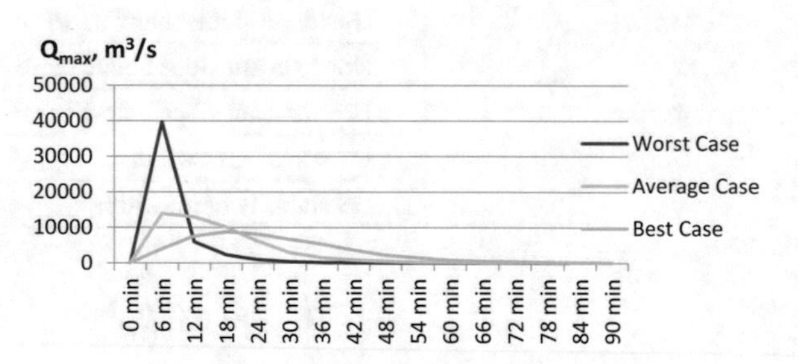

Fig. 14 Geghi dam breach hydrographs

where V_T is total impounded volume (million m^3), V_F is the volume of tailings that could potentially be released (million m^3), D_{max} is the distance to which the material may travel in a downstream channel (run-out distance, km), H_f is predictor.

Total impounded volume in Geghanoush TSF is 4.6 million m^3. Thus, according to the equations above, volume of tailings that could potentially be released (V_F) is 1.4 million m^3 and run-out distance is 15.4 km.

5 Armenian Field Test Cases

The field exercises in selected pilot areas of Armenia, near the borders with Iran, Azerbaijan, and Georgia had several main aims:

To enhance preparedness, response and resilience capacities of the communities in emergencies in accordance with the existing Disaster Risk Management/Emergency Plans through establishment of public-private partnerships. To test and improve the existing contingency plans (or create new ones in the case of Akhtala), based on realistic worst case scenarios, with the active involvement of private sector, NGOs and local communities.

The goal was to assess limitations of rural areas dealing with large, complex emergencies. Rural zones are located far away (in distance and time) from the capital and other major cities, thus the exercises aim was improving the prevention, preparedness and management measures. Increased resilience of local communities was planned to be improved by preparedness capacity building of governing bodies/structures/organizations engaged in the community population protection system and testing of relevant activities.

To evaluate the cross border cooperation in cases where an official agreement exists (case of Akhtala, between Armenia and Georgia), and in cases where there is no agreement but unofficially a strong cooperation exists (the case of Armenian city Kapan, and Iran). The case of Azerbaijan was not examined as no diplomatic relations exist yet.

In all three cases the working team visited the pilot areas many times to gain a detailed knowledge of the existing situation, to establish trust relationships with all key players, and evaluate and agree with them a common picture of the existing conditions and the aims of the exercises and the steps forward. In all pilot areas, activities such as small exercises have taken place, however not all major stakeholders participated or were aware of the results and no capitalization measures were in place.

5.1 *The Road to the Field Exercises*

Prior to field exercises, a series of preparatory activities took place to maximize the benefits of exercises. Table top exercises—one per pilot area—defined the details of the field exercises and finalized the aims, objectives and future steps after exercises execution and evaluation. Training activities were done during table top exercises and prior to field ones to ensure that all key stakeholders their obligations by the existing legislation. This also served to inform citizens in pilot areas on planned exercises and request their participation. Through these activities, existing gaps in plans, procedures and resources were identified. These were combined with results from past events and were recorded and all regional and local partners that should participate in the exercises were mapped.

The ALTER team had already identified good practices that Armenian authorities could benefit if adopted. Such good practices varied from ways of effective engagement of local communities to disaster preparedness and management activities, to webGIS and decision support tools (i.e. evacuation tools, early warning systems). The series of activities, from pilot studies to table top and field exercises helped Armenian authorities to access good practices and decide which ones will be adapted and adopted to help exploitation of field exercises findings.

A dedicated conference in Yerevan summarized all progress and the expansion of "ALTER project" model for improving disaster reduction and management capabilities in mining and dams' management sectors, to other important ones in Armenia.

5.2 *Scenarios*

The main events used for the pilot areas in Armenia are described in brief in the following text covering the scenarios without mentioning the long list of activities that took place to cover all major aspects related to effective disaster preparedness and management.

5.2.1 29 October 2019—Sisian Municipality (Targets: Town of Sisian, Villages Shaghat, Balak, Uyts and Vorotnavan) Participants 326

Emergency Description

In accordance with the developed scenario, a strong earthquake was reported in Sisian Community located in Syunik Province (known as Marz) of Armenia causing damage to the dam of Spandaryan Reservoir. Local authorities were given a certain time for organization of population protection activities in pursuant to the available Disaster Risk Management (DRM) Plan before failure of the dam.

5.2.2 30 October 2019—Kapan City Participants 387

Emergency Description

In accordance with the developed scenario, a strong earthquake was reported in Kapan Community located in Syunik Province of Armenia. The dam of Geghi Reservoir collapsed right after the strong earthquake hit. The local authorities have strict time limits for organization of population protection activities in accordance with available DRM Plan. Autonomy in action the first critical hours, and cross border cooperation with Iran were also considered.

5.2.3 1 November 2019—Akhtala Community. Participants 87

Emergency Description

A massive earthquake is reported in Akhtala Community located in Lori Province of Armenia. There are dozens of completely/partially collapsed buildings and structures due to the earthquake. A collapse occurred at Shamlugh copper mine of Akhtala Mining and Processing Enterprise. As a result of the collapse, 3 miners are left in ruins. There is a threat of failure of the second dam of "Nahatak" tailings dump in consequence the area at risk will be polluted and the road leading to Chochkan village will be blocked. Checking also the procedures for timely and effective activation of cross border cooperation agreement with Georgia.

5.3 Evaluation of the Field Exercises

The Field exercises planned and conducted under the umbrella and coordination of Armenian Government (Ministry of Emergencies (MES) and National Platform for Disaster Risk Reduction (ARNAP)), based on extreme yet realistic scenarios to provide a clear unquestionable picture of existing situation in disaster preparedness and management in rural areas. Moreover, the field exercises provided the framework under which all isolated activities of previous years (small drills, training, informational campaigns, etc.) were evaluated and integrated into disaster risk reduction and management national, regional and local plans. Some of these are:

- Installation of real time warning system sensor packages at the private company ZCMC's premises for providing alerts to Kapan and Sisian in case of dam break and subsequent flood, and installation of early warning sirens systems donated by United Nations and but never used.
- Signing of 3 MOUs between the ALTER project target community local authorities, MES, private entities and ARNAP, based on previous year's isolated efforts and activities by the prementioned entities.
- Evaluation of cross border Armenia—Georgia cooperation agreement, that previous drills had revealed its weaknesses without any improvement steps to be taken.
- Evaluation meetings and procedures after Field Exercises, revealed a number of issues raised require urgent solutions/regulations.

Cross border cooperation agreements depend on strong political willingness (Armenia—Georgia) but are not coupled with joint operating procedures, risk assessment and common pool of resources, neither are tested regularly under the auspices of National Civil Protection Authorities. In other cases, (Armenia—Iran), the absence of official agreements underlines the limits of the unofficial cooperation regional and local authorities have established near the border areas for cases of emergencies.

Local self-governing bodies are not yet in the position to independently accomplish all the Disaster Risk Management (DRM) task procedures. In this respect the legal field needs to be improved.

The role of emergency logistics is crucial for increasing functionality of civil protection authorities the first critical hours after a major disaster and for supporting operations and recovery, yet are not used to optimize the civil protection system. Indicative examples: Depots with crucial for operations equipment are located into flooded zone, transportation alternatives are not considered even though geomorphology of pilot areas call for it (many, deep valleys under landslide risk from earthquake and floods).

Community services and private entities' action plans are not yet harmonized with the community and regional plans. The same applies for the action plans of International Organizations active in rural areas of Armenia.

Private sectors' involvement in the process of developing/implementing the Community DRM plans is considered essential, but there is a need of relevant mechanisms to move forward. Know-how in safety and security from international companies with expertise provided to private companies in Armenia must be diffused into state's disaster risk reduction and management plans and procedures.

Social contributions (Corporate Social Responsibility) to the communities made by private companies should include components of resilience and DRM. Although private entrepreneurship in the mining sector schedules and implements social-promoting measures, the social vector does not directly incorporate the resilience/DRM component. It is necessary to reconsider the guidance of social sector investments and direct a certain portion of the community contributions to the solution of DRM sectorial problems. Appropriate legislative changes to the concept of public-private partnership are needed.

The level of awareness and preparedness of the community population is low. Population training activities are missing even though in the cases of limited time for evacuation, most citizens must act themselves without support of the authorities. Moreover, long term trust building campaigns are missing, even though are necessary in cases with limited time to react, where citizens must act independently of the authorities for a certain time while trusting authorities' activities 100%.

There is a need for clear distinction between the population protection and civil protection issues (civil alerts and emergency alerts are often considered as the same).

There is a need for regular simulation of "Alert—Evacuation" exercise with involvement of all stakeholders and potential resources at the community level. Particularly, there is a need to regularly review the mapping of safe evacuation routes.

During the Field Exercises, almost all evacuations from the buildings were carried out through one familiar exit, although in the buildings there were other available exits with entrance/exit doors. This is just one example of many, showing that results of typical drills that do not really offer to improve the civil protection system. A 2nd example is the fact that in case of sudden flood due dam breaking in Sisian, the fire brigade will manage to save only 50% of their equipment—necessary to protect the population—while the operational center may be under water.

Many of the communities have no alert systems to notify the population during emergencies or natural disasters. In general, the water reservoirs and tailings dumps are not equipped with automatic alert devices. This is a major problem for possible worst case scenarios.

5.4 Outcomes of the Field Exercises

The outcomes after the three field exercises were accepted by the local community and led to:

- Official improvements of the existing contingency plans in Sisian and Kapan, approved by the Ministry of Emergencies (MES), few months after the exercises.
- Created a solid contingency plan in the case of Akhtala, immediately after the exercise, approved by MES.
- Strengthened the existing cooperation between public authorities, private companies, local communities and international organizations in disaster risk reduction and management, through official agreements signed few months after the exercises.
- Revealed the weaknesses in disaster management in rural and cross border areas, initiating an Armenian Government procedure to radically improve the situation.
- The role of emergency logistics proved crucial, not only for supporting operations and offering relief after a disaster but as a key factor to optimize civil protection system. Reallocation of depots and operational centers.
- Led to re-examination of cross border agreement with Georgia to include—in the following 1–2 years—joint risk assessment in the border area, cross border standard operating procedures, commonly planned realistic scenarios and testing them through frequently organize exercises, develop joint training programs and common campaigns for raising awareness on risks.

All initiatives bring to the Armenian authorities background to start a procedure for improving legislation and operational plans and procedures, based on good practices tested in field exercises and their findings.

6 Desktop Open Source GIS for Flood Hazard Map Visualization

Based on the flood risk mapping simulation results for Kapan area a specially developed open source desktop tool has been created. Its architecture was based on model designed within the ALTER consortia with no use of commercial software. The application is developed using open source GIS software; the server part for the dynamic events, was created by JavaScript and its libraries and frameworks. The tools implemented were open source software solutions such as: Geoserver, Qgis, Web App Builder, Boundless WEBSDK, Open Layers.

Geoserver allows the user to display spatial information to the world;

QGIS is a professional GIS (Geographic Information System) cross-platform application that is Free and Open Source Software (FOSS);

Web App Builder is a plugin for QGIS that allows easy creation of web applications;

Boundless WEBSDK which provides tools for easy-to-build JavaScript-based web mapping applications;

OpenLayers is an open-source JavaScript library for displaying map data in web browsers.

It includes different features and tools that may lead to faster response and easier way of taking decisions in flood event cases.

The application has the function to visualize the most vulnerable buildings (Fig. 15). It includes different scenarios that can be analyzed in operational room and by its tools can support better management of the current and future situation in cases of flood events. It is focused on visualization of high waves coming after dam break in cases of failure.

The application has the ability of switching the predefined layers and the base map layers. The predefined layers display very rich data by turning them off or on. Users can easily make analysis of the risks in cases of flood events. It includes different scenarios of the water spread in support of better decision making and faster resource allocation. Layers can be downloaded as geojson files. Geolocation of team members on the field is available for the users.

The base map layers are including Street map, Satellite map, Shaded relief map and NatGeo map which can be used in operational room analysis.

A draw feature tool can mark the zone of interest (Fig. 16) by polygon or line which will be visualized and be seen in the operational room in real time.

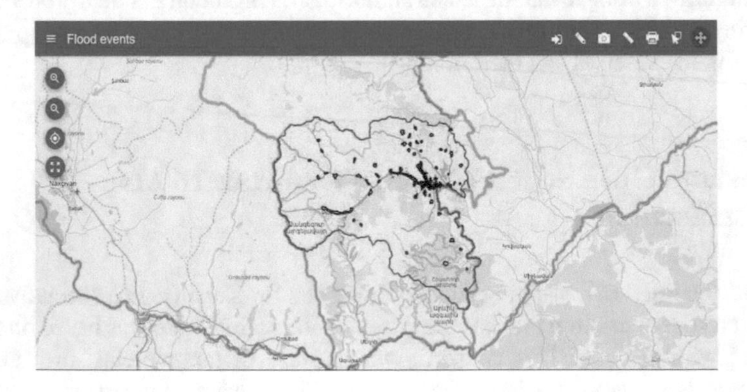

Fig. 15 Application main screen

Fig. 16 Layers list

Fig. 17 Application Popup

The Popup feature visualizes information about the vulnerable buildings such as: schools, kindergartens and others (Fig. 17).

The export feature can save maps with new data as picture format files. This feature can be used in future data analysis.

Measure and distance options can be used to measure the distances and also can measure the size of the focused area.

The application provides connection to the current weather forecast via open weather with detailed information about the current or future weather conditions. It is connected to EFAS emergency management service which provide extra satellite data about current conditions.

7 Conclusion

The analyses of ALTER project were very important for the Armenian stakeholders in the pilot communities. The exercises conducted after the profound dam break analyses and the creation of the potential flood risk maps was of great importance for the Armenian emergency management authorities. Such initiatives can also be implemented in the optimal control of the mine procedures [13].

Additionally, the skills and technology transferred by the project will remain for future use in order to be built upon with new capabilities. The webGIS system and server installed will serve as the hosting platform for further applications in disaster management and other fields. Additionally, seismic monitoring stations installed on a previously unmonitored dam will serve as a best practice for dam and reservoir monitoring throughout Armenia.

Finally, the connections and relationships between ALTER partners, researchers, and ministry officials in Armenia represent a strong link for continued best practices in monitoring and risk management moving forward.

Acknowledgements This work has been supported by the DG ECHO project called: "Alliance for disaster Risk Reduction in Armenia" with acronym: ALTER with Grand Reference: 783214 and the Bulgarian National Scientific Fund project number DFNI DN12/5 "Efficient Stochastic Methods and Algorithms for Large-Scale Problems".

References

1. Georisk, C.J.S.C.: Assessment of the Multi-Component Risk Determined by the Maximum Seismic Impact on the Kapan City (Multi-Hazard City Scenario). Project # ARM 10-0000005849. Final Report (2017)
2. Gevorgyan, A., Minasyan, R., Khondkaryan, V., Antonyan, A.: The Prediction of Possible Flooding of the Territory as a Result of the Accident of the Geghi Reservoir Dam (2014)
3. Wahl, T.L.: Dam breach modeling—an overview of analysis methods. In: Joint Federal Interagency Conference on Sedimentation and Hydrologic Modeling, 27 June–1 July 2010, Las Vegas, NV
4. Using HEC-RAS for Dam Break Studies: Compiled by Hydrologic Engineering Center of US Army Corps (2014)
5. Concha Larrauri, P., Lall, U.: Tailings dams failures: updated statistical model for discharge volume and runout. Environments **5**, 28 (2018)
6. Rico, M., Benito, G., Diez-Herrero, A.: Floods from tailings dam failures. J. Hazard. Mater. **154**, 79–87 (2008)
7. Chambers, D.M., Bowker, L.N.: Tailings Dam Failures 1915–2017. Available online: https://www.csp2.org/tsf-failures-1915-2017. Accessed on 16 Aug 2017
8. SCS: Simplified Dam-Breach Routing Procedure. United States Department of Agriculture, Soil Conservation Service (SCS). Technical Release No. 66 (Rev. 1), 39 (1981)
9. Wahl, T.L.: Prediction of Embankment Dam Breach Parameters—A Literature Review and Needs Assessment. Dam Safety Research Report, DSO-98-004. Water Resources Research Laboratory, U.S. Dept. of the Interior, Bureau of Reclamation, Dam Safety Office (DSO), July 1998

10. Wahl, T.L.: Evaluation of new models for simulating embankment dam breach. In: Association of State Dam Safety Officials (ASDO) Conference, Hollywood, Florida, 27 Sept–1 Oct 2009
11. Schlaffer, S., Harutyunyan, A.: Working Paper: LCLU Voghji River Basin. AUA Acopian Center for the Environment, AUA GIS and Remote Sensing Lab (2018)
12. Saaty, T.L.: Multicriteria Decision Making: The Analytic Hierarchy Process: Planning, Priority Setting, Resource Allocation (1990)
13. Traneva, V., Tranev, S.: Existence of a solution of the problem of optimal control of mines for minerals. Adv. Stud. Contemp. Math. **21**(3) (2018)

Fire Simulator Capable to Analyze Fire Spread in Real Time with Limited Field Weather Data. Case Study—Kresna Fire (2017)

Nina Dobrinkova and Adrián Cardil

Abstract Bulgaria is not the most wildfire vulnerable country on the Balkan Peninsula. However the country exposure to wildfires in the last twenty years is increasing constantly. Wildfires are becoming larger and their duration is more than one or two days as it was before the year 2000. Bulgarian authorities monitoring the forest publish every year statistics which are freely available and confirm that worrying trend. In this situation proactive measures have been taken by international team from Bulgarian Academy of Sciences and León, Spain testing a new tool called Wildfire Analyst (WFA). That software can provide real-time analysis of wildfire behavior and simulates the spread of wildfires fast using only personal computer. The tests done with one of the largest fires in Bulgaria in the last 5 years (Kresna 2017) were quite promising. Thus the team decided to present the first findings and use them for future work in the most vulnerable south situated wildfire prone Bulgarian zones.

Keywords WFA tool · Wild fire simulations in real time · Kresna fire 2017 · Calibration

1 Introduction

Historically wildfires in Bulgaria start to be monitored in more extensive way since 1970 (Fig. 1) [1]. In 2006 the Ministry of agriculture, forest and food transfer this monitoring activities to the Executive Forest Agency of Bulgaria (Figs. 2 and 3) [2]. Both authorities provide annually reports stating that the fire causes are mainly based on carelessness from human activities in wildlands. The information in this reports clearly state that Bulgaria is having increasing number of affected territories with

N. Dobrinkova (✉)
Institute of Information and Communication Technologies, Bulgarian Academy of Sciences, acad. Georgi Bonchev bl. 2, 1113 Sofia, Bulgaria
e-mail: ninabox2002@gmail.com

A. Cardil
Tecnosylva. Parque Tecnológico de León, 24009 León, Spain
e-mail: acardil@tecnosylva.com

S. Fidanova (ed.), *Recent Advances in Computational Optimization*, Studies in Computational Intelligence 920,
https://doi.org/10.1007/978-3-030-58884-7_2

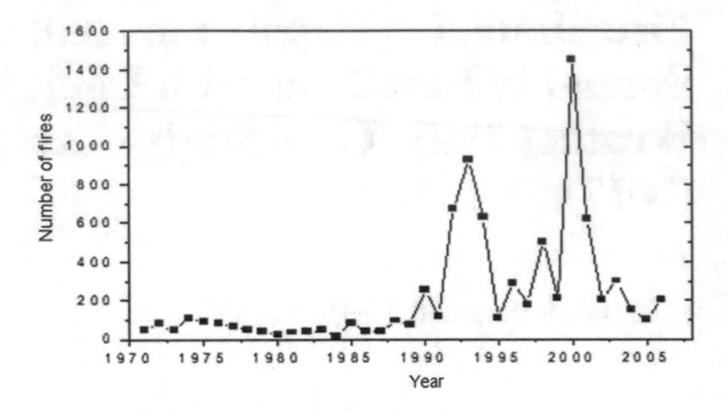

Fig. 1 Number of wildfires in Bulgaria 1971–2006 [1]

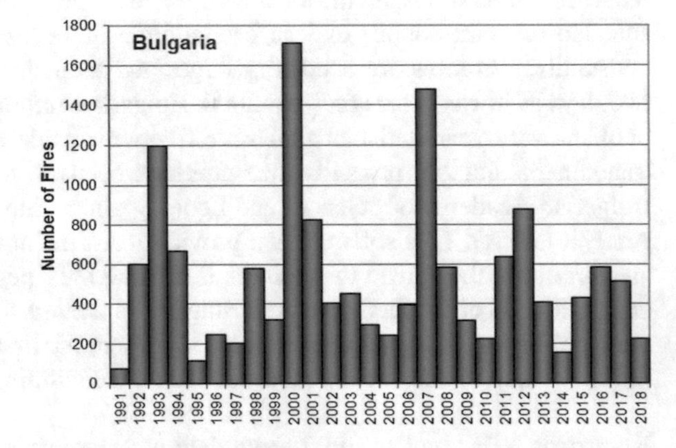

Fig. 2 Number of wildfires for Bulgaria in the period 1991–2018 [2]

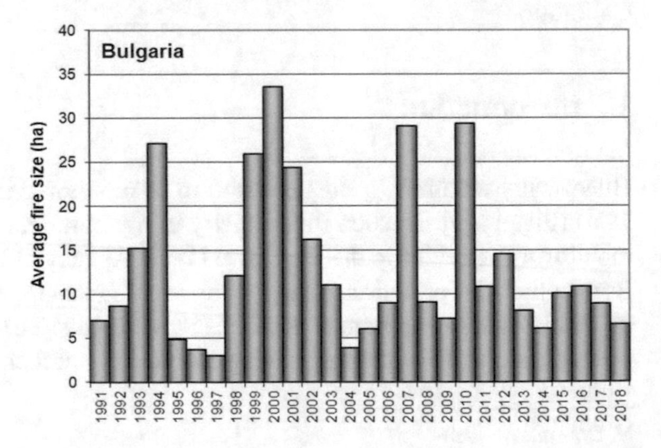

Fig. 3 Average wildfire size in (ha) for Bulgaria in the period 1991–2018 [2]

wildland fires after the year 1990. Since that year the number of fires increase drastically having repetition peaks every 6–8 years. Most probably because the vulnerable areas understory regrowth again and the harmful human activities remain unchanged during that period.

Currently, in Bulgaria, computer-based wildfire analysis tools are rarely used to support firefighters and incident commanders on field activities. Commonly, historical data is collected and used to reconstruct past fire events on a paper. Usually the work is limited to running fire simulations in scientific laboratories, but no further implementation is elaborated by operational teams. Some satellite images are used for the analysis of the fire perimeter and affected zones as well as input data to feed the fire spread models. In our article we will focus on a new tool called Wildfire Analyst (WFA) applied for Bulgarian test case, where weather data from the field was missing. The tool is a fire simulation framework that provides near real-time analysis of wildfire progression, fire behavior, suppression capabilities, and impact analysis for an incident. WFA has been developed to be used in fire operations by implementation of algorithms that allows doing the fire progression calculations in near-real time with user friendly interface. The software has the option to incorporate data from GIS layers done outside the tool and use in it as direct inputs for simulation. Assignment of Fire Behavior Fuel Models (FBFMs) can be done directly in the attributive table of each GIS layer that contain biological information of what is burning on ground. The so called burning materials could be divided by the user implementing either the 13 or the 40 known classes or creation of custom fuel models is also possible. Meteorological data is extracted inside of the tool getting direct information by the available meteorological services for the zone or if available local collected data sets can be inserted. Adjustment mode during the process of simulation is available for the wildfire analyst specialist in the cases fire progression is not realistic to the field information.

This new tool is relatively young, but is already used in North and South America, as well as in Spain. All results presented for this three locations are quite positive and that was the reason testing and calibrations have been performed for Bulgaria's biggest fire in the last five years (Kresna Gorge Fire 2017). We have selected the test fire for the simulator by evaluating the fire database in the ministry of Interior. The chosen fire was active between 24 and 29 August 2017 and burned 1600 ha in the first run and extra 1600 ha by the end of its activity phase. All preliminary data collection and preparation for simulation process will be summarized in this paper.

The team goal was to analyze the accuracy of the fire behavior outputs provided by Wildfire Analyst (WFA; [3]) in Bulgaria by using the Kresna fire as a case study. EU reports [4] showed wildland fires are more frequent in the last decades, especially in South EU countries. Particularly, officially published statistics from Bulgarian responsible authorities presents figures about burned area in all affected areas in Bulgaria, which are in line with the EU trend (Figs. 1, 2 and 3 [1, 2]). Fire regime is changing and fire behavior and spread is getting more extreme from 1970s to nowadays.

After the year of 2000 wildland fires are frequent and the fire season is getting larger and having two windows to have large fires: (1) early spring: March and April; (2) summer: from July to October. In 2017 wildland fires affected Southern parts of Bulgaria, even in areas where fire occurrence is not usual. One of these areas was Kresna Gorge where is located Struma Motorway as part of the EU corridor IV connecting Vidin (North-west Bulgaria) and Thessaloniki (North-west Greece) areas. Kresna Gorge fire spread in the period 24–29 August 2017 and caused damages larger than 15 million leva (approx. 7.5 million EUR), burned 7 houses, 1 car and 3200 ha, including large Natura 2000 protected zones. During the fire, the Trans-European corridor IV was closed and not operational, triggering high economic losses for transport companies and local population. In our paper we will evaluate the application of sophisticated computer based tools like WFA in the decision support process of the local authorities and decision makers and how this software could support their work for operational needs.

2 Wildfire Simulation Tools Applied Since 2009 for Bulgarian Test Cases

Experimental wildfire simulations for Bulgaria started in 2009 for the territory of Harmanli region by application of the WRF-Fire (SFIRE) fire spread model [5, 6]. The first fire simulations with the coupled atmospheric model were performed for ideal cases in order to be analyzed the accuracy and model way of working under different variations of meteorological and terrain data for the selected zone in south Bulgaria. The first study area selected was Zheleznitsa village which is located in Vitosha Mountain near by Sofia city district. The results from these first simulations provided a good framework to know how the model can be initialized and what input data was useful to model fire spread in Bulgaria by using a real wildfire event. Later, the real case study was used to calibrate and adjust the fire spread model for southeast Bulgarian test zone. This real fire was selected from the national fire data base of the Ministry of agriculture and forests in 2010. The fire was active in the period 14–17 August 2009. The fire modeling was focused on two domains including the ignition sources in the villages of Ivanovo, Leshnikovo and Cherna Mogila, near by Harmanly. This area was located in South-East Bulgaria close to the Bulgarian-Greek border. The topography and fuel used for the simulations at 60 m resolution had difference with the geogrid program in the core of the simulator, because of two 2 extra variables—NFUEL_CAT and ZSF. NFUEL_CAT. They were set in order to introduce to the simulator the 13th Anderson Fire Behavior Fuel models [7] as land cover. The best simulation results in terms of simulated burned area are shown in Figs. 4 and 5.

UC Denver super computer was used for running the fire spread models and estimate the fire progression shown in Fig. 4. Figure 5 was the real fire spread in the nature.

The next step to implement fire simulators and modelling fire spread for Bulgarian active fires was done through the analysis of a new case study. It was located to the territory of Zlatograd forestry department located in the south central Bulgaria. Both BEHAVE Plus and FARSITE fire simulators were used to model the fire spread. The Fire Behavior Fuel Models (FBFMs) applied to this case study were the 13th Anderson models and the 40th models introduced by Scott-Burgan in 2005 [7, 8].

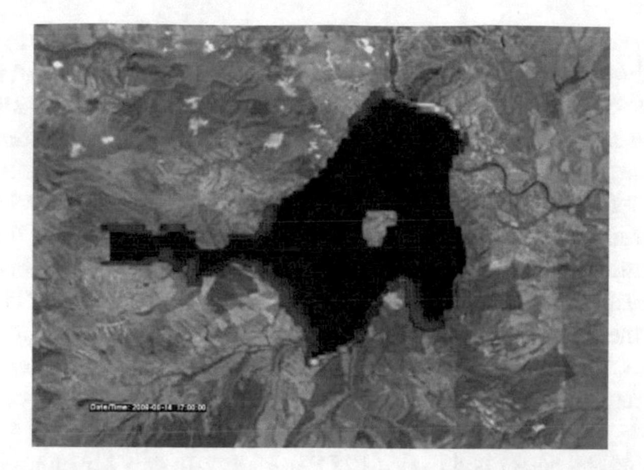

Fig. 4 The simulation fire burnt area

Fig. 5 The real fire burnt area

Fifteen wildfires occurred in 2011 to 2012 on the territory of Zlatograd forestry department. They were analyzed after collecting weather and fuel data from the field. Based on the initial BehavePlus outputs using standard fuel models, custom fuel models were developed for some vegetation types that were not well represented by the US fuel models in this Bulgarian site.

After the static fire behavior analysis performed through BehavePlus and the calibration of fuel models, FARSITE was used to dynamically model the fire spread, defining test landscapes using a 500 m buffer zone around each of the fifteen Zlatograd fires.

Several spatial inputs are required in FARSITE to perform single fire simulations, including topography, vegetation, and fuel parameters compiled into a multi-layered "landscape file" format. Canopy cover values were visually estimated from orthophoto images and verified with stand data from the Zlatograd forestry department. Tabular weather and wind files for FARSITE were compiled using weather and wind data from TV Met, a Bulgarian meteorological company that provided hourly records. Tabular fuel moisture files were created using the fine dead fuel moisture values calculated by using BehavePlus analyses (1-h timelag fuels). The 10-h fuel moisture content value was estimated by adding 1% to the 1-h fuel moisture content and the 100-h fuel moisture was generally calculated by adding 3% to the 1-h fuel moisture. The live fuel moisture values previously estimated from BehavePlus weather scenarios were used in terms of live herbaceous and live woody moisture values.

All simulations performed in FARSITE used metric data for inputs and outputs. Adjustment fuel value was not used to modify the rate of spread for standard fuel models; rather custom fuel models were created. Crown fire, embers from torching trees, and growth from spot fires were not enabled.

An example of one of the successful FARSITE runs is a fire that occurred on August 30, 2011, starting at 1400 and ending around 1800, and burned a total area of 0.3 ha. The input parameters to this small grassland fire in FARSITE were set as follows:

Fuel moisture values: 6% (1-h), 7% (10-h), 9% (100-h), 45% (live herbaceous), and 75% (live woody);

Daily maximum temperatures: 17–21 °C;

Daily minimum relative humidity: 24–50%;

Winds: generally from the west-southwest at 1–2 km/h.

The fire size as calculated using FARSITE was 0.5 ha, which could be the case if the suppression measures of the people could be included in the system representation. The results from the simulated and real burnt areas are presented in Fig. 6.

More information about Behave Plus and FARSITE simulators used for the Zlatograd test cases can be found in [9].

3 Wildfire Analyst (WFA) Tool

After having a retrospective analyses how the fire spread simulations evolve for the Bulgaria rest cases the new available tool called Wildfire Analyst is the next in our list for evaluation. This software provides real-time analysis of wildfire behavior and simulates the spread of wildfires (Fig. 7). Simulations are completed quickly, in seconds, to support real time decision making. WFA was specifically designed to address this issue, providing analysis capabilities for a range of situations and users. WFA provides seamless integration with ESRI's ArcGIS with no conversion or pre-processing required. This increases its usability, allowing users to concentrate

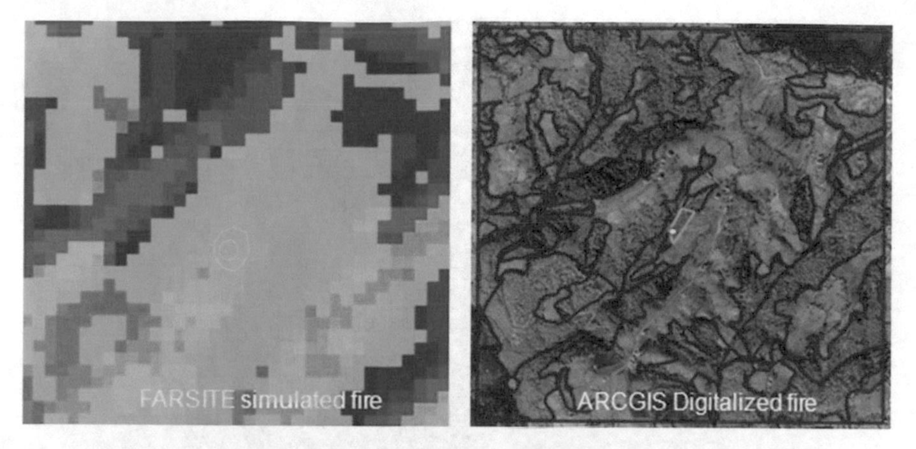

Fig. 6 FARSITE run for a grassland fire, difference in shape similarity in size

Fig. 7 Simulation of a fire in the US with wildfire analyst

on interpreting simulation outputs, and making important decisions about how and where to deploy firefighting resources. This software was designed to be used with a laptop or tablet at the incident command center, in the operations center, or directly on scene, providing outputs in less than a minute (Fig. 8). The software can use predefined weather scenarios, or current and forecasted weather obtained via web services, to model fire behavior and provide outputs within seconds. For wildland fire, time is of the essence, and WFA was specifically architected to support initial attack situations, giving the Fire Chief and Incident Commander the critical intelligence needed to support resource allocation decision making. This fire simulator tool provides a range of analytical outputs, available as GIS maps and charts that empower more accurate and timely decision making. The desktop platform, or web

Fig. 8 Simulation's results visualized on laptop, PC or tablets

and mobile enabled applications, capabilities and results are easy to use. WFA is a software component of Technosylva's incident management software suite. WFA provides seamless integration with fiResponse™ platform (Technosylva), to provide real time information about where a fire is likely to spread and what potential impacts may be—in seconds. Wildfire Analyst provides critical information to support the resource dispatching and allocation.

Contemporary fire behavior software tools required a high degree of specialization, training and effort in the preparation and conversion of GIS data to use the software. Historically, this has been a limitation in using these tools for initial attack and real time applications. This fast performance facilitates use of the outputs in real time and allows for constant adjustment based on field observations and deployment decisions by the incident team. It is a powerful tool in the arsenal of the incident team, providing simulations very quickly, and repeatedly as changing conditions dictates.

WFA can use different fire spread models like Rothermel (1972) [10] or Kitral (1998) [11] that is generally used in South America. Additionally, it has embedded other models such as Nelson equations to estimate fuel moistures from weather data (2000) [12] or WindNinja [13], a software that computes spatially varying wind fields for wildland fire applications in complex terrain. Input data required by the software is similar to other fire simulators such FARSITE or BehavePlus due to it uses the same fire spread models and equations.

Also, WFA introduces new simulation modes with innovative enhancements including real time processing performance, automatic rate of spread (ROS) adjustments based on observations used to create fire behavior databases, calculation of evacuation time zones (or 'firesheds'), and integration of simulation results for asset and economic impact analysis. It has been used operationally for diverse agencies worldwide. For instance, Military Emergency Unit and several regions in Spain, CONAF and private companies in Chile, Italy or US wildfire services are using this tool.

4 Simulation Results for Kresna Gorge Fire (2017) Test Case

The data available for the Kresna Gorge fire is fragmented in different Bulgarian agencies and not combined in one single source. Thus the team which was working on data collection and preparation for its usage under the WFA simulation combined efforts from Bulgarian Academy of Sciences and Forestry University in Sofia.

As a first step we had to estimate the area of the fire, which has as starting point Kresna Gorge, located north of Stara Kresna village. The fire spread was in direction south to the village of Vlahi. In the Western direction, the fire had descended on a steep slope almost to the Struma River, but there was a lower intensity of the fire spread. In this area north winds are very probable to happen, so local weather conditions created by the fire could be the case for the time frame when it was active.

The general coordinates of the fire can be described as follows:

- Starting Region as ignition point and its North Border (Approx.): 41.818961°; 23.179666° (for the first date of its occurrence 24th Aug. 2017)
- Western border: 41.814869°; 23.160966°
- Eastern border for the first two days of the fire spread: 41.801595°; 23.220162°
- Eastern border for the last days: 41.777043°; 23.224270° (this are the coordinates of the fire in its end phase on 29th Aug. 2017)
- Southern limit: 41.746235°; 23.234691°

As next steps was to start data collection of the local meteorological conditions, relief and fuel load.

During the period of the fire occurrence on 24th August 2017 and up until its end on 29th August 2017 all meteorological data from the local meteorological stations was elaborated and provided for WFA simulations in predefined format with one hour time step by the National Institute of Meteorology and Hydrology in Bulgarian Academy of Sciences.

The local relief was used from raster 30 × 30 DEM information available for Bulgaria about the area of interest.

Fire Behavior Fuel Models (FBFMs) were used by applying the Anderson [7] and Scott-Burgan [8] in order to obtain the burning materials. This data was extracted from forestry maps provided by the local unit of the ministry of agriculture, forest and food of Bulgaria.

The burned area captured by the local authority's satellite measurements can be seen on Fig. 9.

Fig. 9 Kresna Gorge burned area one day before fire end [14]

The Kresna test case had a hilly and low-mountain terrain, situated at elevations from about 250 m a.s.l. to 750 m a.s.l.. The fire affected the eastern slopes of Kresna Gorge above Struma River and roughly between the villages Mechkul (to the north) and Vlahi (to the south).

The fire started in grassland-shrub zone and transferred to plantations from Pinus nigra, where it quickly spread assisted by the dry conditions. The total affected area was about 2260 ha, of which 65% were forest territories. The fire burnt mostly plantations from Austrian pine (870 ha), Scots pine (Pinus sylvestris), Robinia pseudoacacia (21.1 ha) (1260 dka), natural forests dominated by Quercus pubescens (200 ha) and Quercus petraea (66 ha) and smaller patches of other species. Pine plantations were very affected by fire with rate of mortality higher than 80% due to a high burn severity. On steeper slopes above Struma River the severity of burning was lower, where some deciduous trees and single specimen of Juniperus excelsa survived. The affected zone is mostly protected one under the NATURA 2000 areas.

The used for calibrations tool with this test case was Wildfire Analyst (WFA), which has the real time prediction abilities as module embedded in its functionalities. WFA introduces new simulation options including real time processing performance, automatic rate of spread (ROS) adjustments based on observations used to create fire behavior databases, calculation of evacuation time zones (or ‘firesheds’), and integration of simulation results for asset and economic impact analysis [15, 16].

Adjustment of fire simulations in real-time by identifying adjustment points (also often mentioned as control points) of the fire based on observations of the fire behavior on the field, namely the position of the fire at a certain point of time is useful functionality in WFA. With these adjustments the simulation results tend to be more reliable during the following hours of fire propagation. On a common basis these adjustments are estimated based on intuition and own experience and usually require a large number of simulations and rehearsals to obtain accurate adjustments of the fire. This frequently makes the adjustment unfeasible for operational purposes due to the need of spending a lot of time. WFA automatically allows the adjustment of the ROS of existing fuels and is given by the following formula:

$$ROS = ADJUSTMENT \times ROS\ (humidity, fuel, wind, slope)$$

where the adjustment is represented by a certain value and the ROS is the rate of spread of the fire which is influenced by several factors during its calculation such as the humidity, fuel, wind, and slope of the terrain.

The initialization of WFA has been done with predefined layers in GIS format. All data was collected from different sources that could give us idea about the fire ignition point, spread and types of fuels which were burning on the field.

For a reference have been used satellite data from VIIRS. This was a way to compare the official reports from the responsible authorities and the real fire spread captured by the satellite images (Fig. 10).

The fuel map generated based on the biological species of Kresna Gorge biodiversity has been generated and a sample is presented to Fig. 11. The runs with WFA of the Kresna fire gave as the best match to the real fire progression the map from Fig. 12.

The obtained results have been performed with very limited weather data inputs because the nearest meteorological stations were outside the zone of the fire. Thus the embedded weather scenarios in WFA were used and gave really close to the reality outcome scenario maps.

5 Conclusions

The tests done with WFA have been performed in a way that all available data for southwest Bulgaria was implemented. However the simulation results could be better if the fire simulator would have incorporated bigger map of Fire Behavior Fuel Models in the tested zone. Other shortcoming to the calibration process was that no meteorological stations were available near by the fire area. Thus most of the weather data from the field was interpolated and the first runs with this calculations were giving not promising results.

The embedded functionality of the WFA tool giving the option to obtain weather data from satellite services located nearby the area of the simulations gave much more promising results and the map generated on Fig. 12 was done as the best match of the team simulations efforts.

As future work is foreseen to be developed bigger Fire Behavior Fuel coverage map for south Bulgaria. Its purpose will be all biological species to be converted to Burnable Fuel Models in a way that in real time any user of WFA for Bulgarian test case at least in south part of the country to be able to simulate fire growth using either

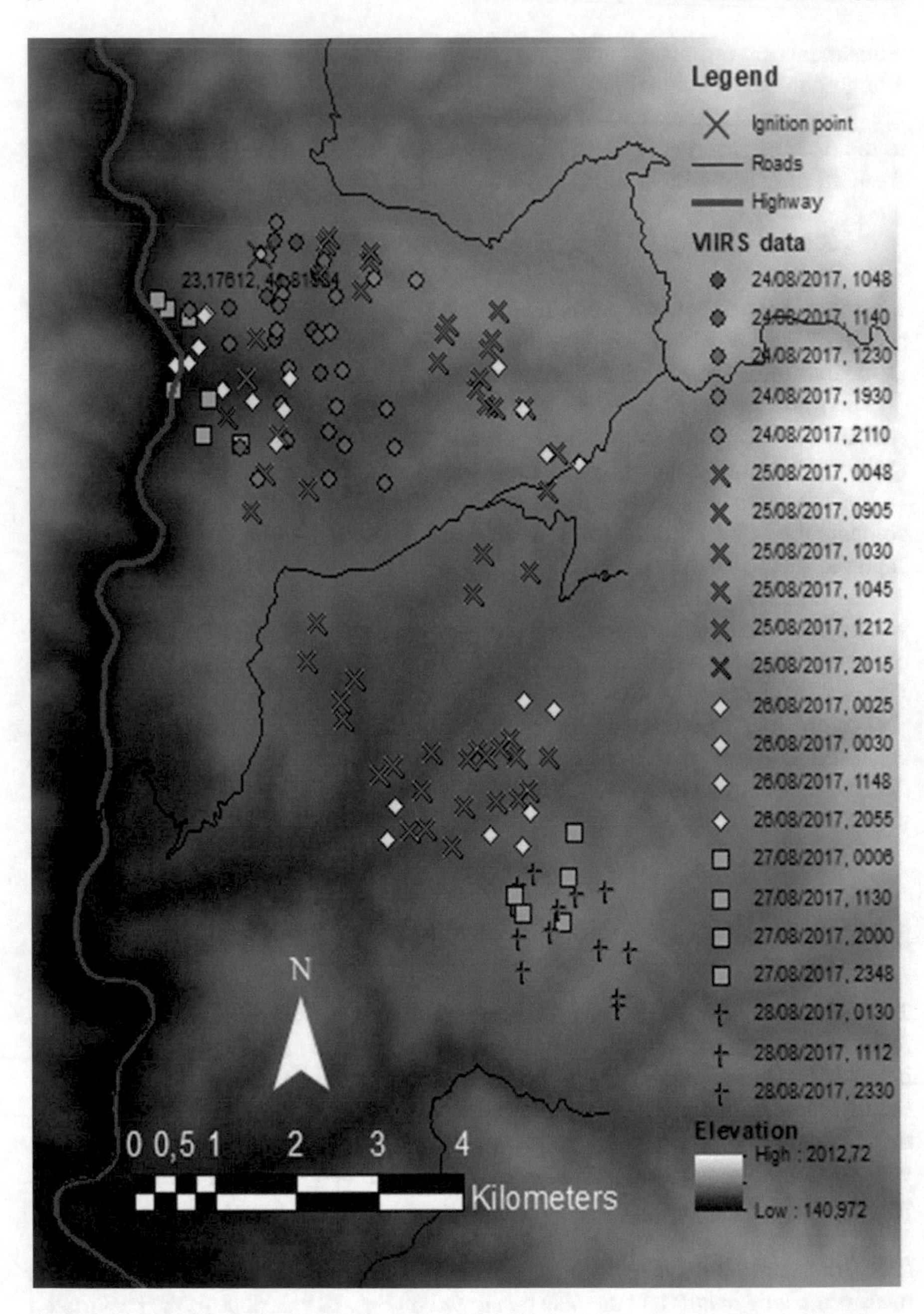

Fig. 10 Kresna fire progression through VIIRS data

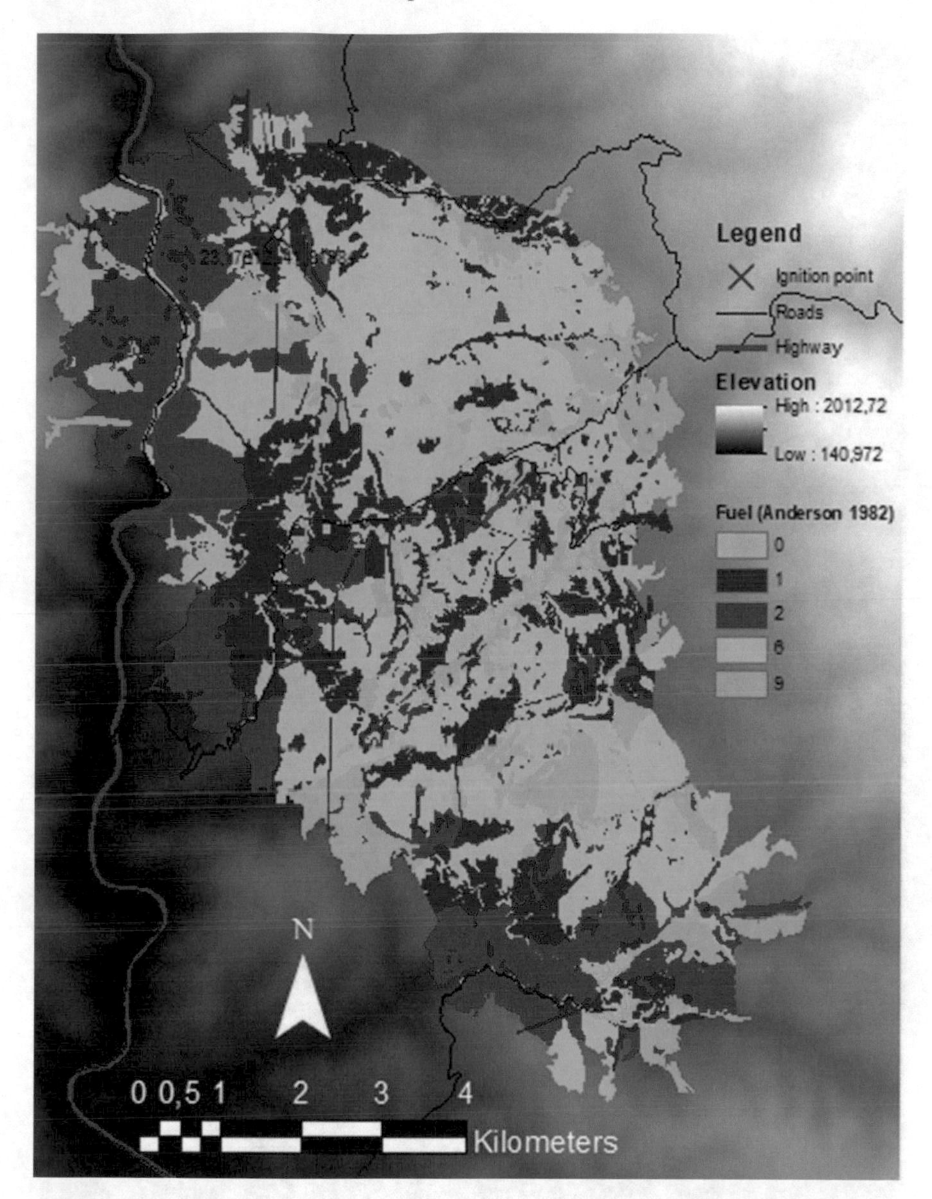

Fig. 11 Kresna fire fuels

nearby located meteorological measurement stations or in case they are not available to use the embedded WFA meteorological options from satellite estimations.

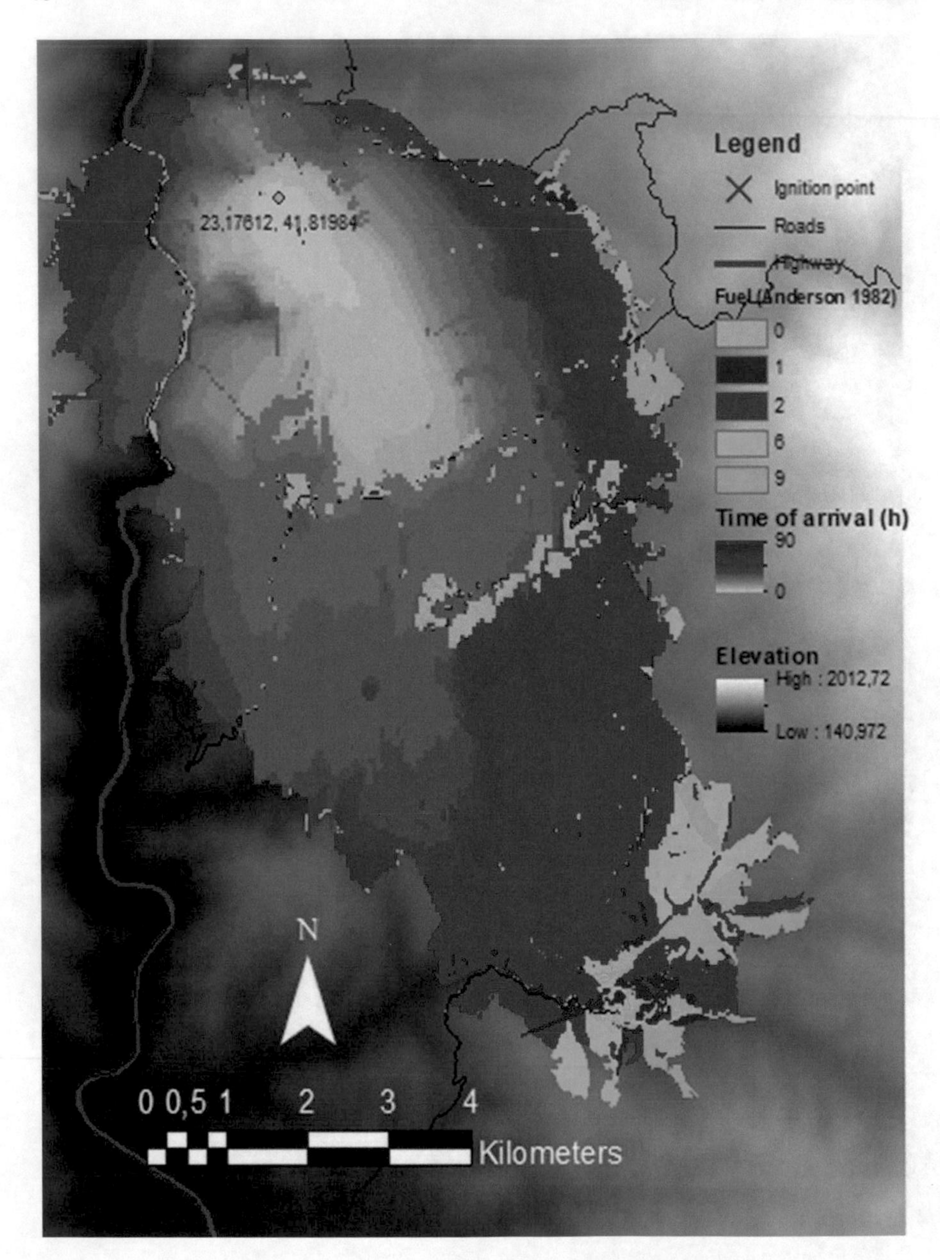

Fig. 12 Kresna fire run with WFA and 1 km buffer zone around the burned area

Acknowledgements This research has been supported by the Bulgarian National Scientific Program "Environmental Protection and Reduction of the Risk of Adverse Events and Natural Disasters" Approved by Council of Ministers Decision No 577/17.08.2018 and funded by the Ministry of Education and Science (Agreement No D01-230/06-12-2018).

References

1. Ecopolis, Bulletin 48: Forest fires reach catastrophic scales (In Bulgarian) (2001)
2. San-Miguel-Ayanz, J., Durrant, T., Boca, R., Libertà, G., Branco, A., De Rigo, D., Ferrari, D., Maianti, P., Artes Vivancos, T., Pfeiffer, H., Loffler, P., Nuijten, D., Leray, T., Jacome Felix Oom, D.: Forest fires in Europe, Middle East and North Africa 2018, EUR 29856 EN. Publications Office of the European Union, Luxembourg, 2019. ISBN 978-92-76-11234-1 (online), 978-92-76-12591-4 (print), https://doi.org/10.2760/1128 (online), https://doi.org/10.2760/561734 (print), JRC117883
3. Cardil, A., Molina, D.M.: Factors Causing Victims of Wildland Fires in Spain (1980–2010). Hum. Ecol. Risk Assess. An Int. J. **21**, 67–80 (2015). https://doi.org/10.1080/10807039.2013.871995
4. Official web-site reports of EFFIS: https://effis.jrc.ec.europa.eu/reports-and-publications/annual-fire-reports
5. Jordanov, G., Beezley, J.D., Dobrinkova, N., Kochanskski, A.K., Mandel, J., Sousedik, B.: Simulation of the 2009 Harmanli fire (Bulgaria). In: 8th International Conference on "Large-Scale Scientific Computations" LSSC'11, Sozopol 6–10 June 2011. Lecture Notes in Computer Science, No. 7116, pp 291–298. ISSN 0302-9743, 2012. Science, Springer Germany (2012)
6. Dobrinkova, N., Jordanov, G., Mandel, J.: WRF-fire applied in Bulgaria. In: Numerical Methods and Applications, 20–24 Aug, Borovez. Lecture Notes in Computer Science, No. 6046. Springer, Germany. ISSN 0302-9743 (2011), pp. 133–140. SJR 0.332. https://doi.org/10.1007/978-3-642-18466-6_15
7. Anderson, H.E.: Aids to determining fuel models for estimating fire behavior. USDA Forest Service, Intermountain Forest and Range Experiment Station, Research Report INT-122 (1982). www.fs.fed.us/rm/pubsint/intgtr122.html
8. Scott, J.H., Burgan, R.E.: Standard fire behavior fuel models: a comprehensive set for use with Rothermel's surface fire spread model. General Technical Report. RMRS-GTR-153. U.S. Department of Agriculture, Fort Collins. Forest Service, Rocky Mountain Research Station. 72 p (2005)
9. Dobrinkova, N., Hollingsworth, L., Heinsch, F.A., Dillon, G., Dobrinkov, G.: Bulgarian fuel models developed for implementation in FARSITE simulations for test cases in Zlatograd area. In: Proceedings of 4th Fire Behavior and Fuels Conference, 18–22 Feb 2013, Raleigh, NC and 1–4 July 2013, St. Petersburg, Russia, pp. 513–521. E-proceeding: https://www.treesearch.fs.fed.us/pubs/46778 (2014)
10. Rothermel, R.: A Mathematical Model for Predicting Fire Spread in Wildland Fuels. USDA Forest Service, Intermountain Forest and Range Experiment Station, (Ogden, UT). Research Paper INT-115 (1972)
11. Pedernera, P., Julio, G.: Improving the economic efficiency of combatting forest fires in Chile: the KITRAL System. USDA Forest Service General Technical Report. PSWGTR-173, pp. 149–155 (1999)
12. Nelson, R.M., Jr.: Prediction of diurnal change in 10-h fuel stick moisture content. Can. J. For. Res. **30**, 1071–1087 (2000). https://doi.org/10.1139/CJFR-30-7-1071
13. Forthofer, J., Shannon, K., Butler, B.: Simulating diurnally driven slope winds with WindNinja (link is external). In: Proceedings of 8th Symposium on Fire and Forest Meteorological Society; 13–15 Oct 2009, Kalispell, MT (2,037 KB; 13 pages) (2009)

14. In Bulgarian local news about Fire spread in Kresna Gorge. https://trud.bg/община-кресна-стартира-дарителска-ка/
15. Ramirez, J., Monedero, S., Buckley, D.: New approaches in fire simulations analysis with wildfire analyst. In: The 5th International Wildland Fire Conference. Sun City, South Africa (2011)
16. Monedero, S., Ramirez, J., Molina-Terrén, D., Cardil, A.: Simulating wildfires backwards in time from the final fire perimeter in point-functional fire models. Environ. Model. Softw. **92**, 163–168 (2017)

Utilizing Minimum Set-Cover Structures with Several Constraints for Knowledge Discovery on Large Literature Databases

Jens Dörpinghaus, Carsten Düing, and Vera Weil

Abstract A lot of problems in natural language processing and knowledge discovery can be interpreted using structures from discrete mathematics. In this paper we will discuss the search query and topic finding problem using a generic context-based approach. This problem can be described as a Minimum Set Cover Problem with several constraints. The goal is to find a minimum covering of documents with the given context for a fixed weight function. The aim of this problem reformulation is a deeper understanding of both the hierarchical problem using union and cut as well as the non-hierarchical problem using the union. We thus choose a modeling using bipartite graphs and suggest a novel reformulation using an integer linear program as well as novel graph-theoretic approaches.

1 Introduction

In scientific research, expert systems provide users with several methods for knowledge discovery. They are widely used to find relevant or novel information. For example, medical and biological researchers try to find molecular pathways, mechanisms within living organisms or special occurrences of drugs or diseases. In [1], we discussed a novel approach for describing NLP problems using theoretical computer science. Using this approach, it is possible to obtain the algorithmic core of a NLP problem. Here, we will discuss two $\mathcal{NP}$-complete problems: Search Query Finding (SQF) and Topic Finding (TF).

Using expert system as an input, researches usually consider an initial idea and some content like papers or other documents. The most common approach is inquir-

J. Dörpinghaus (✉) · C. Düing
Fraunhofer Institute for Algorithms and Scientific Computing, Schloss Birlinghoven, Sankt Augustin, Germany
e-mail: jens.doerpinghaus@scai.fraunhofer.de

V. Weil
Department for Computer Science, University of Cologne, Cologne, Germany
e-mail: weil@informatik.uni-koeln.de

S. Fidanova (ed.), *Recent Advances in Computational Optimization*,
Studies in Computational Intelligence 920,
https://doi.org/10.1007/978-3-030-58884-7_3

ing a search engine to find closely related information. Thus two question are most frequently asked: "How can I find these documents?" to adjust the search query for knowledge discovery or "What are these documents all about?" to find the topic. Both questions are heavily related to the context of documents. Meta-data like authors, keywords and text are used to retrieve results of a query using a search engine.

Semantic searches are usually based on textual data and some meta-data like authors, journals, keywords. In addition, time and complexity play an important role, since often relevant information is not findable or new information is already available. For example, databases such as PubMed [2] contain around 27 million abstracts and PMC[1] includes around 2 million biomedical-related full-text articles.

Especially in the field of biomedical sciences, there is a long history of developing applications that assist researchers. For instance, SCAIView[2] is an information retrieval system that allows semantic searches in large textual collections by combining free text searches with the ontological representations of automatic recognized biological entities (see Hodapp et al. [3]). SCAIView was used in many recent research projects, for example regarding neurodegenerative diseases [4] or brain imaging features [5]. Furthermore, it was also used for document classification and clustering [6]. Another important real-world task is the creation of biological knowledge graphs that is tackled by the BELIEF environment [7]. It assists researchers during the curation process by providing relationships extracted by automatic text mining solutions and represented in a human-readable form [8]. At the core of both technologies several implementations of the methods of biomedical text mining are in place. See Fig. 1.

Both problems—Search Query Finding (SQF) and Topic Finding (TF)—are equivalent (see [1]) and can be described as a Minimum Set Cover Problem with several constraints. Query languages and natural languages are not only highly connected but merge more and more (see [9] or [10]). The goal is to find a minimum covering of documents with the given context for a fixed weight function. The aim of this problem reformulation is a deeper understanding of both the hierarchical as well as the non-hierarchical problem. We thus choose a modeling using bipartite graphs and suggest a novel reformulation using an integer linear program as well as graph-theoretic approaches.

There is a considerable amount of literature on both problems. Many studies have been published on probabilistic or machine-learning-approaches, see [11, 12] or [13]. In addition, in recent years there has been growing interest in providing users with suggestions for more specific or related search queries, see [14].

This paper is divided into six sections. Section 2 gives a brief overview of the problem formulation and provides the definition of MDC and WMDC. Section 3 analyses the hierarchical problem formulation and proposes novel heuristics. In Sect. 4, we present a short analysis of the non-hierarchical problem and propose an integer lin-

[1] https://www.ncbi.nlm.nih.gov/pmc/.

[2] https://www.scaiview.com/ (an academia version is freely available at http://academia.scaiview.com/academia/).

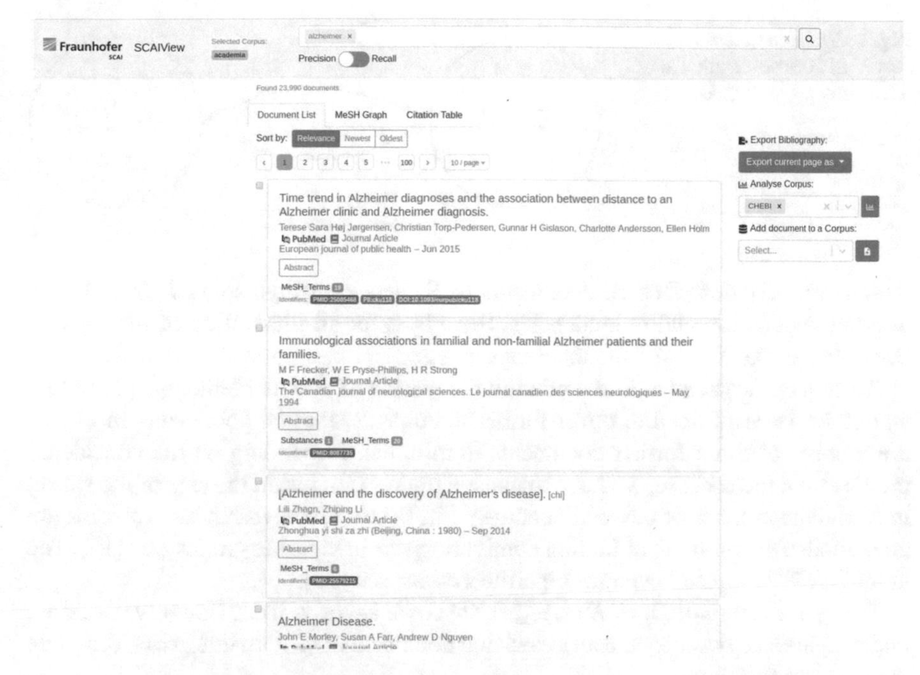

Fig. 1 Screenshot of SCAIView showing the result list for free text search “alzheimer”. This is usually a starting point for both Topic Finding (TF). If the corpus is explicitly given, this would be a starting point for Search Query Finding (SQF)

ear program approach and some modified graph heuristics to solve this problem. We present some experimental results on artificial and real-world scenarios in Sect. 5. Our conclusions are drawn in Sect. 6.

2 Problem Formulation and Definition

We follow the notation introduced in [1]. Let $\mathbb{D}$ be a set of documents and let $\mathbb{X}$ be a set of context data. Context data is information associated with documents, such as keywords, authors, publication venue, etc. Both $\mathbb{D}$ and $\mathbb{X}$ form the vertex set of a graph G. If and only if a description of a document $d \in \mathbb{D}$ is associated with context data $x \in \mathbb{X}$, we add the edge $\{d, x\}$ to E. The graph $G = (\mathbb{D} \cup \mathbb{X}, E)$ is bipartite and called *document description graph*.

Given a subset $R \subset \mathbb{D}$, the search-query-finding (SQF) or topic-finding (TF) problem tries to find a good description of R with terms in $\mathbb{X}$. In general, we lack a proper definition of what *good* means.

For example, given a search engine $q : \mathbb{X} \to \mathbb{D}$ and a description function $f : \mathbb{D} \to \mathbb{X}$, we want a solution $Z \subset \mathbb{X}$ such that $q(Z) = R$ and $Z = f(R)$. If we want to obtain a human-readable topic for R, we need a solution Z of minimum cardinality

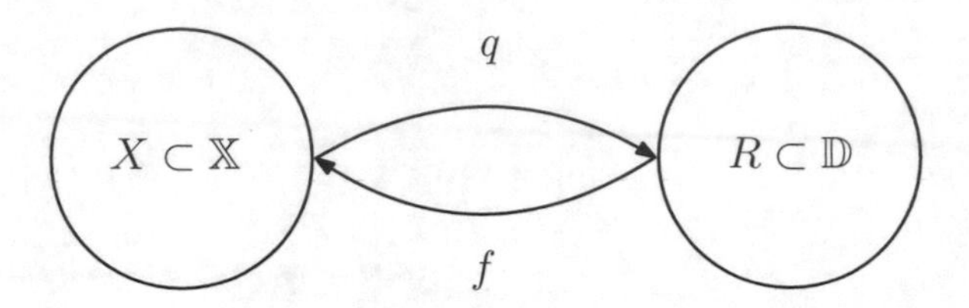

Fig. 2 Relation between the sets $X \subset \mathbb{X}$ as description set of documents in $R \subset \mathbb{D}$.

which precisely describes all documents in R, hence distinguishing R from $\mathbb{D} \setminus R$ without duplication and redundancies. See Fig. 2 for an illustration of the relation between the sets X, R and the mappings f, q.

To sum up, we need to find a minimum covering of R with elements in $\mathbb{X}$ so that whenever we are forced to cover further documents, that is, documents in $\mathbb{D} \setminus R$, the number of these further documents is minimal. Depending on the considered problem and the usecase, we have to make a trade-off between the size of the subset in $\mathbb{X}$ and the number of covered documents in $\mathbb{D} \setminus R$. However, these problems are all related to the problem of finding dominating sets in bipartite graphs, see [15]. The latter is $\mathcal{NP}$-complete, even for bipartite graphs, see [16].

For $x_i \in \mathbb{X}$, we call $D_i = N(x_i) \subseteq \mathbb{D}$ the cover set of x_i in $\mathbb{D}$. Roughly speaking, just imagine a keyword x_i and all associated documents D_i. With this, we reformulate the problem as follows:

Definition 2.1 (*Document Cover Problem, DC*) Let $\mathbb{D}$ be a set of documents, let $\mathbb{X}$ be a set of context data and let $G = (\mathbb{D} \cup \mathbb{X}, E)$ be the document description graph.

Given a set of documents $R \subset \mathbb{D}$, a solution of the DC is a set $C \subseteq D$ that covers at least R. □

Definition 2.2 (*Minimum Document Cover Problem, MDC*) Let C be a solution of the DC and let $\alpha_2 = |C|$. Let further $\alpha_1 = r$ be the number of documents in $C \setminus R$.

A solution of MDC is a solution of DC so that $\alpha = \alpha_1 + \alpha_2$ is minimal. □

We can define two objectives for minimization: α_1 and α_2.

Definition 2.3 (*α_2-Minimum Document Cover Problem, α_2-MDC*) Given a set of documents $R \subset \mathbb{D}$, a solution of the α_2-MDC is a solution of DC so that $\alpha = \alpha_1$ is minimal. □

Definition 2.4 (*α_1-Minimum Document Cover Problem, α_1-MDC*) Given a set of documents $R \subset \mathbb{D}$, a solution of the α_1-MDC is a solution C of DC so that $\alpha = \alpha_2$ is minimal. □

We further introduce a weighted version of this problem:

Definition 2.5 (*Weighted Minimum Document Cover Problem, WMDC*) Let $\mathbb{D}$ be a set of documents, let $\mathbb{X}$ be a set of context data and let $G = (\mathbb{D} \cup \mathbb{X}, E)$ the document description graph. Let $w : \mathbb{X} \to \mathbb{R}$ be a weight function which associates a weight for every element in $\mathbb{X}$. Moreover, we set $D = \{D_1, \ldots, D_n\}$. Let $\alpha_1 = r$ be the number of documents in $R \subset \mathbb{D}$ and $\alpha_2 = |C|$.

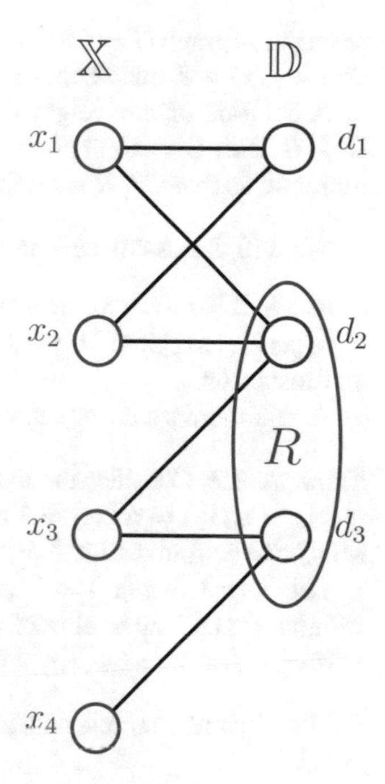

Fig. 3 A graph $G = (\mathbb{D} \cup \mathbb{X}, E)$ illustrating Example 2.8

A solution of the WMDC is a set $C \subseteq D$ which covers R, such that $\alpha = \alpha_1 + \alpha_2 + w(C)$ is minimal, where $w(C) = \sum_{c \in C} w(c)$. □

Again we can find formulations for α_1-WMDC and α_2-WMDC. Both problems are $\mathcal{NP}$-hard, see [17].

In general, we will focus on the α_2 optimization. Thus, in this paper, we denote this version with the MDC and WMDC.

We have to distinguish between hierarchical and non-hierarchical approaches. Both MDC and WMDC search for a cover set $c_1, \ldots, c_n$ which leads to a solution $c_1 \cup \cdots \cup c_n$. This is a non-hierarchical approach. Using a search engine this would lead to a solution c_1 `or ... or` c_n. Utilizing the cut of sets we will need a hierarchical solution $(c_1 \cup \cdots c_n) \cap (c_{n+1} \cup \cdots c_m) \cap \cdots$. Using a search engine would lead to a solution (c_1 `or ... or` c_n) `and` (c_{n+1} `or ... or` c_m)) `and`….

Definition 2.6 (*Hierarchical Minimum Document-Cover Problem, HMDC*) Let $\mathbb{D}$ be a set of documents, let $\mathbb{X}$ be a set of context data and let $G = (\mathbb{D} \cup \mathbb{X}, E)$ be the document description graph. Moreover, we set $D = \{D_1, \ldots, D_n\}$.

A solution of the HMDC problem for $R \subset \mathbb{D}$ is a minimum cover $C \subseteq D$ with $C = C_1 \cap \cdots \cap C_n$ and $C_i = C_1^i \cup \cdots \cup C_m^i$ of R so that $C \setminus R$ is minimal. We use $N(x_i)$ as usual for the open neighborhood $N(x_i) \setminus x_i$. □

Definition 2.7 (*Hierarchical Weighted Minimum Document-Cover Problem, HWMDC*) Given a set of documents $\mathbb{D}$, a set of context data $\mathbb{X}$ and the document description graph $G = (\mathbb{D} \cup \mathbb{X}, E)$. We set $D = \{D_1, \ldots, D_n\}$. Given a weight function $w : \mathbb{X} \to \mathbb{R}$ that defines a weight for every element in $\mathbb{X}$.

A solution of the weighted HWMDC problem for $R \subset \mathbb{D}$ is a minimum cover $C \subseteq D$ with $C = C_1 \cap \cdots \cap C_n$ and $C_i = C_1^i \cup \cdots \cup C_m^i$ of R, i.e. $\sum_{c \in C} w(c)$ is minimal, so that $C \setminus R$ is minimal. □

We will discuss two examples for the non-hierarchical problem:

Example 2.8 Given an instance of the MDC with $\mathbb{D} = \{d_1, d_2, d_3\}$, $R = \{d_2, d_3\}$, $\mathbb{X} = \{x_1, \ldots, x_4\}$ and $D_1 = D_2 = \{d_1, d_2\}$, $D_3 = \{d_2, d_3\}$, $D_4 = \{d_3\}$. See Fig. 3 for an illustration.

A minimum set cover cannot include x_1 or x_2, but a solution is $C = D_3$. □

Example 2.9 Consider the instance given in Example 2.8 with additional weights $w(x_1) = w(x_2) = w(x_3) = 1$ and $w(x_4) = 0$. A minimum solution of the weighted MDC can be found with $Z = \{x_3, x_4\}$.

Let $w(x_1) = w(x_3) = 1$ and $w(x_4) = w(x_2) = 0$. A minimum solution of weighted MDC can be either found with $Z = \{x_2, x_4\}$, here $w(Z) = 0$ but $|C \setminus R| = 1$. If we chose $Z = \{x_3, x_4\}$ $w(Z) = 1$ but $|C \setminus R| = 0$. □

We will now continue with a formal definition according to the notation introduced in [1].

2.1 Generating and Optimisation of Search Queries

In this paper we use a very generic definition of search engines and search queries. A search engine is a function $q : \mathbb{X} \to \mathbb{D}$ which outputs a set of documents or any other content of the domain set if the input is a subset of a description set $\mathbb{X}$ which we call search query.

The problem of generating search queries (SQF) usually has a domain set $\mathbb{D}$ restricted by the database of the search engine. The return value of our problem is a search query $\mu \in \mathbb{X}$ so that $q(\mu) = R$. Thus R is the subset of documents for which we want to create a search query.

We have a mapping from one element in $\mathbb{X}$ to a subset of $\mathbb{D}$. q is thus the right inverse of the description function f and $f \circ q = \mathrm{id}_{\mathbb{X}}$. Not only does f have to be surjective, but we also have to assume that even q is surjective. Every document in the target set $\mathbb{D}$ should be a target of some search query.

It is very easy to see that this is usually not given in reality: Assume q is a websearch, $\mathbb{X}$ the web search description and $\mathbb{D}$ the set of all web pages available. Some of them may not be indexed due to restrictions made to the robots crawling and indexing the web. We can sail around this by restricting $\mathbb{D}$ to $q(\mathbb{X})$. Then f should be the right inverse of q with $q \circ f = \mathrm{id}_{\mathbb{D}}$.

We can also see, that $\forall d \in \mathbb{D}$ several $\mu_1, \ldots, \mu_n \in \mathbb{X}$ exists with $d \in q(\mu_i)$—neither q nor f are injective mappings. If we want to find the optimal μ we need to define some sort of metric on elements in $\mathbb{X}$. This can be very complex. If we assume, that we have a terminology T and a simple algebra with $\vee$ and $\wedge$, we can simplify $\mathbb{X} = \mathcal{P}(T, \vee, \wedge)$ and take the length of $\mu \in \mathbb{X}$ as a metric. But if all documents have a unique index stored in $\mathbb{X}$ the shortest search query might consist of a concatenation of these indexes listing all documents in R.

Thus, the simplest evaluation function $e : \mathbb{D} \rightarrow [0, 1]$ is set by

$$err_1(d_i, d_j) = \begin{cases} 1 & i \neq j, f(d_i) \neq f(d_j), d_i, d_j \in R \\ 1 & i \neq j, f(d_i) = f(d_j), d_i \text{ or } d_j \notin R \\ 0 & \text{else} \end{cases} \tag{1}$$

If f, the description function with the image set $\mathbb{X}$, does not map two documents in R to the same element, which is the search query $\mu \in \mathbb{X}$, we count an error. Same happens, if another document not in R is mapped to μ. Thus we want to find a description function f so that $f(R) = \mu \in \mathbb{X}$ with $q(\mu) = R$. It follows that the problem is given by

$$p = \mathbb{D}|R|\mathbb{X}|err_1|R$$

This is the simplest formulation of the stated problem. As discussed, it can be more complex. We have not defined a proper quality measure for search queries $\mu \in \mathbb{X}$. In addition, the space $\mathbb{X}$ may be very complex and it is not clear, if it is—like $\mathbb{D}$—a discrete space with a proper metric. In addition, although f is a surjective mapping and q can be set to be surjective, it is left open, if one of these mapping might also be injective.

2.2 Complexity

To analyze the complexity of this problem, we will first of all focus on the easiest version. Let $\mathbb{X} = \mathbb{N}$ be the set of identifiers for documents. We have a bijective function $doc : \mathbb{X} \rightarrow \mathbb{D}$ that maps an identifier to a document. The inverse function is $doc^{-1} = id : \mathbb{D} \rightarrow \mathbb{X}$. Then a search query $\mu \subseteq \mathbb{X}$ is defined as a list of n identifiers $\mu = [\mu_1, \ldots, \mu_n]$. A search query is clearly defined as $q(\mu) = [doc(\mu_1), doc(\mu_2), \ldots]$. Then solving the Search Query Problem can be done in polynomial time by computing $f(R) = [id(d_1), \ldots, id(d_m)]$ for $R = \{d_1, \ldots, d_m\} \subseteq \mathbb{D}$.

Going one step forward, we may consider more than one identifier. Let's assume we have n meta data fields which are not unique. For example, we may have `journal:"The New York Times"` or `author:"Einstein, Albert"`. Then $\mathbb{X} = F \times L$ is the space of all n meta data field F and their data L. We have a function $docs : \mathbb{X} \rightarrow \mathbb{D}$ which returns for a given element $x \in \mathbb{X}$ all documents containing this meta data. Let us assume that our search query is denoted by $q(\mu) = [docs(\mu_1) \cup docs(\mu_2) \cup \ldots]$. The only logic considered by this is `and`.

Solving the Search Query Problem can be done by solving the set cover problem. We have a finite, non empty set $\mathbb{D}$ and a family $F = \{docs(\mu_1), docs(\mu_2), \ldots\}$ of subsets of $\mathbb{D}$ for all $\mu_i \in \mathbb{X}$. We want to find a subset $G \subset F$ which covers $R \subset \mathbb{D}$ with

$$\bigcup_{S_i \in G} S_i = R$$

There is no need to add weights to these sets, because we are searching for a minimum set cover of R which already meets error measure 1. This problem is $\mathcal{NP}$-complete as shown by Garey and Johnson [18]. In addition it is part of Karp's 21 NP-complete problems list, see [19].

Before continuing, we will propose an exact definition of the SQF-Problem, assuming that both functions *docs* and q can be solved polynomial time.

Definition 2.10 (*Search-Query-Finding Problem (SQF)*) Let $\mathbb{D} = \{d_1, d_2, \ldots\}$ be a nonempty set of documents, $R \subseteq \mathbb{D}$ with $\emptyset \neq R$ and $S_1, S_2, \ldots \subseteq \mathbb{D}$ nonempty subsets. Let $\mathbb{X}$ be a nonempty description set and $\mu \subseteq \mathbb{X}$, $\mu = \{\mu_1, \mu_2, \ldots\}$ a search query. Further we need the mappings $docs : \mathbb{X} \to \mathbb{D}$ and search engine $q : \mathbb{X} \to \mathbb{D}$, with $docs(\mu_i) = S_i$ and $q(\mu) = [docs(\mu_1) \cup docs(\mu_2) \cup \cdots]$. The function q ist the rightinvers of the description function $f : \mathbb{D} \to \mathbb{X}$. Let $F = \{docs(\mu_1), docs(\mu_2), \ldots\}$ be a family of subsets of $\mathbb{D}$ for all $\mu_i \in \mathbb{X}$. □

Now we will define the optimization version and the corresponding decision version of the SQF:

Optimization Problem: Finding a minimum sized $G \subseteq F$ which covers $R \subseteq \mathbb{D}$ with

$$\bigcup_{S_i \in G} S_i = R.$$

That means we need to find a minimum sized $\mu \subseteq \mathbb{X}$, such that $q(\mu)$ covers R. This is what we will call the SQF, because we are interested in a minimum cover.

Decision Problem: Given an integer k, we want to know, if there is such a subset $G \subseteq F$ that covers $R \subseteq \mathbb{D}$, with $|G| \leq k$, which implies $|\mu| \leq k$. It will be named as D-SQF for short.

Lemma 2.11 *The SQF-Problem is solvable in polynomial time, if there is a unique identifier* id *and a bijective function* $doc : \mathbb{X} \to \mathbb{D}$ *exists that maps an identifier to a document.* □

Lemma 2.12 *The SQF-Problem is* $\mathcal{NP}$*-hard, if there is no unique identifier* id *which means no bijective function* $doc : \mathbb{X} \to \mathbb{D}$ *exists that maps an identifier to a document.* □

Starting with the decision problem we will prove the $\mathcal{NP}$-completeness of the D-SQF-Problem. Therefor we first need to proof, that D-SQF is in $\mathcal{NP}$.

Lemma 2.13 *The D-SQF-Problem is in* $\mathcal{NP}$. □

Proof We guess a subset $\mu \subseteq \mathbb{X}$ and check, if $q(\mu) = G$ is a cover of R and if $|\mu| \leq k$. This can be done in polynomial time. □

Lemma 2.14 *The D-SQF-Problem is $\mathcal{NP}$-hard.* □

Proof We need to show, that SET COVER $\leq_p$ D- SQF.

Given an instance of SET COVER, we construct a instance of D- SQF. An instance of SET COVER is given by a set $U = \{x_1, x_2, \ldots, x_n\}$, a collection of m subsets $M_i \subseteq U$, an integer k and a collection C of at most k of the subsets M_i such that their union covers U.

We define the instance of D- SQF just by identifying $U = R$, $\{x_1, x_2, \ldots, x_n\} = \{d_1, d_2, \ldots, d_n\}$, $M_i = S_i$ and $C = G$. μ follows as $\mu = q^{-1}(G) = \{docs^{-1}(\mu_1) \cup docs^{-1}(\mu_2) \cup \cdots\} = f(G)$. All this can be done in polynomial time. □

Now Lemma 2.12 directly follows from Lemmas 2.14 and 2.15 if the optimization version is polynimial equivalent to the decision version.

Lemma 2.15 *The* SQF-*Problem is polynomial equivalent to the* D- SQF. □

Proof $\Leftarrow$: Let $\mathbb{D} = \{d_1, d_2, \ldots\}$ be a nonempty set of documents, $R \subseteq \mathbb{D}$ with $\emptyset \neq R$ and $S_1, S_2, \ldots \subseteq \mathbb{D}$ nonempty subsets. Let $\mathbb{X}$ be a nonempty description set and $\mu \subseteq \mathbb{X}$, $\mu = \{\mu_1, \mu_2, \ldots\}$ a search query. Further we need the mappings $docs : \mathbb{X} \rightarrow \mathbb{D}$ and search engine $q : \mathbb{X} \rightarrow \mathbb{D}$, with $docs(\mu_i) = S_i$ and $q(\mu) = [docs(\mu_1) \cup docs(\mu_2) \cup \cdots]$. The function q ist the rightinvers of the description function $f : \mathbb{D} \rightarrow \mathbb{X}$. Let $F = \{docs(\mu_1), docs(\mu_2), \ldots\}$ be a family of subsets of $\mathbb{D}$ for all $\mu_i \in \mathbb{X}$.

Compute for a vector $k \in \{0, 1\}^{|X|}$ each permutation $\sigma(k)$ and the return value of $q(\sigma(\mathbb{X})$. With this we can find a minimal k^* so that the decision version answers with yes. This can be done in linear time according to the size of $\mathbb{X}$ and over all in polynomial time if q can be computed in polynomial time.

$\Rightarrow$: Let $\mathbb{D} = \{d_1, d_2, \ldots\}$ be a nonempty set of documents, $R \subseteq \mathbb{D}$ with $\emptyset \neq R$ and $S_1, S_2, \ldots \subseteq \mathbb{D}$ nonempty subsets. Let $\mathbb{X}$ be a nonempty description set and $\mu \subseteq \mathbb{X}$, $\mu = \{\mu_1, \mu_2, \ldots\}$ a search query. Further we need the mappings $docs : \mathbb{X} \rightarrow \mathbb{D}$ and search engine $q : \mathbb{X} \rightarrow \mathbb{D}$, with $docs(\mu_i) = S_i$ and $q(\mu) = [docs(\mu_1) \cup docs(\mu_2) \cup \cdots]$. The function q ist the rightinvers of the description function $f : \mathbb{D} \rightarrow \mathbb{X}$. Let $F = \{docs(\mu_1), docs(\mu_2), \ldots\}$ be a family of subsets of $\mathbb{D}$ for all $\mu_i \in \mathbb{X}$.

Given an optimal subset $X \subset \mathbb{X}$ so that $q(X)$ is an optimal solution with value k^* of the SQF-Problem. Consider for each element $x \in X$ all elements x' with $q(x') \subset q(x)$. Let k be the value of the result of $q(X')$ which uses x' instead of x. For each this implies a solution of the decision version with value $k \geq k^*$ and the answer will be yes. This can be done in quadratic runime. □

Consequently it is an important step to either apply well-known approximations or heuristics to solve the underlying set-covering problem. Not need to say, that common search engines have a much higher complexity. It is easy to add full text or keyword search. Applying logical operators is much more complex, since it would add cuts to the set-cover problem. Nevertheless, since usually a list of identifiers is not a result people want so see, we can assume, that the SQF-Problem is usually $\mathcal{NP}$-hard.

2.3 Generating and Optimisation of Cluster Labels

A clustering is usually done on a domain set $\mathbb{D}$ and leads to several clusters $C_1, \dots, C_n, n \in \mathbb{N}$. If $\mathbb{D} = \mathcal{P}(\mathbb{S})$, these clusters are explicitly coded in the set $\mathbb{D}$. Finding cluster labels (CLF) is the task of assigning a subset of a description set $\mathbb{X}$ with the description function $f : \mathbb{D} \to \mathbb{X}$ to a cluster $R \in \{C_1, \dots, C_n\}$. We might consider an evaluation function measuring the distance between the description between two documents in R, $|f(d_i) - f(d_j)|$. But we need to assume a proper metric on $\mathbb{X}$ to do so. This leads to very complex questions. For example: What is a proper metric on a space of boolean algebra? The easiest evaluation function is thus given by

$$err_2(d_i, d_j) = \begin{cases} 1 & i \neq j, f(d_i) \neq f(d_j), d_i, d_j \in R \\ 1 & i \neq j, f(d_i) = f(d_j), d_i \text{ or } d_j \in R \\ 0 & \text{else} \end{cases}$$

Here we define that every two documents in R must share the same cluster labels. This cluster label has to be unique to this cluster. The reference standard can also be set to R. Thus the problem of generating and optimisation of cluster labels is given by

$$p = \mathbb{D}|R|\mathbf{X}|err_2|R$$

where the resulting label set is the image $f(R) \subset X$. Depending on the choice of X this either leads to a set of metadata, terms, sentences or any subset of natural language. Again, this problem can be very complex.

2.4 Complexity

We can apply the results obtained in Sect. 2.2. We already discussed, that Hagen et al. found out that both problems—SQF and CLF—are similar, see [20]. It is easy to proof that given the same domain set $\mathbb{D}$, image set $\mathbf{X}$ of the description function and the same evaluation function both problems are equivalent. Thus, they are in general closely connected.

Lemma 2.16 *Let $\mathbb{X}$ be a description image set. For every solution f of $p_1 = \mathbb{D}|R|\mathbf{X}|err_1|R$ this is also an optimal solution of $p_2 = \mathbb{D}|R|\mathbf{X}|err_2|R$.* □

Proof This follows directly, since $err_1 = err_2$. □

Same follows directly for the inverse:

Lemma 2.17 *Let X be a description image set. For every solution f of $p_2 = \mathbb{D}|R|\mathbf{X}|err_2|R$ this is also an optimal solution of $p_1 = \mathbb{D}|R|\mathbf{X}|err_1|R$.* □

Thus both problems are equivalent if we consider the same domain set $\mathbb{D}$, image set X of the description function and the same evaluation function. In the next section we will discuss that this is usually not the case. But for some special application both problems can be solved in the same way. The level of abstraction between natural language and discrete structures brings the most challenging computational problems. But we need to state the following lemma:

Lemma 2.18 *CLF is NP-hard.* □

We can conclude that we can use the same or similar heuristics for solving both problems. Usually a search query language is not used for representing cluster labels. But query languages and natural languages are not only highly connected but merge more and more (see [9] or [10]).

We will first of all focus on hierarchical approaches, discussing approaches using dynamic programming and bipartite graph heuristics or spanning trees. After that we will discuss the non-hicrarchical problem and solutions using an integer linear program approach as well as some heuristics utilizing the graph structure. We will evaluate the results on some random instances and finish with a conclusion.

3 Hierarchical Approaches

3.1 Problem Description

For some questions it is interesting to find a cover of $R \subset \mathbb{D}$ with increasing (decreasing) or selectable exactness and the number of named entities $Z \subset X = f(R)$. If we have a set of documents and want to obtain more others closely related documents, we may be interested in a modification of the similarity measure for documents or search queries. We build covers $C_i = q(Z_i)$ of R and optimize the solution by concatenating them with a logical AND.

3.2 Using Unique Keyword Descriptions on Bipartite Graphs

From the graph in Fig. 3 we can see that the graph $G = (\mathbb{D} \cup \mathbb{X}, E)$ is bipartite. The neighborhood $N(d) \subset \mathbb{X}$ of every document $d \in \mathbb{D}$ is not necessarily unique description of this document. Thus we can find a trivial solution of the MDCP on $R \subset \mathbb{D}$ by

$$\vee_{d \in R}(\wedge_{x \in N(d)} x)$$

We can eliminate elements with the largest error from this list. This process can be limited by iterations as well as a precision. For example we may limit the precision to 0.9 which will eliminate at maximum 10% of all keywords, whereas a precision of 0.5 will eliminate at maximum 50%.

Algorithm 1 KEYWORD-COVER

Require: Documents $\{d_1, \ldots, d_n\} \subset \mathbb{D}$ and descriptive elements $f(d_i) = \{x_1, \ldots, x_m\} \subset \mathbb{X}$, a weight function $w : \mathbb{X} \rightarrow \mathbb{R}$ maxiter as maximum of iterations, prec as precision
Ensure: A cover $Z = (x_i \wedge x_j \wedge \ldots) \vee (x_k \wedge x_l \wedge \ldots) \vee \ldots$ of R with elements in $\mathbb{X}$.

```
   f' = f
2: for every d ∈ R do
     while iteration<maxiter AND f'(d) > (prec · f(d)) do
4:     remove x ∈ f'(d) with maximum weight
     end while
6: end for
   return Z = ∨_{d∈R}(∧_{x∈f'(d)} x)
```

If we set $w : \mathbb{X} \rightarrow \mathbb{R}$ as the error function $err(x) = |q(x) \setminus R|$ we will find a solution for MDCP, otherwise this will return a solution of WMDCP. The function err is a less time-consuming approach but highly depended on the distribution of $\mathbb{X}$.

3.3 *Dynamic Programming and Bipartite Graph Heuristic*

Here, we describe a heuristic and dynamic method by creating dominating subgraphs of a bipartite graph. Building the bipartite graph $G_b = (V = R \cup X, E)$, a subgraph of the document description graph $G = (\mathbb{D} \cup \mathbb{X}, E)$, we create a set with documents $R_a = \{d_1, \ldots, d_n\} \subseteq \mathbb{D}$ and all their context data (like keywords, named entities, etc.) in a sorted list $X_a = \{x_1, \ldots, x_m\} \subseteq \mathbb{X}$ for the two sets of nodes. The edges (d_i, x_j) in G_b are given for all pairs d_i, x_j iff $x_j \in f(d_i)$. The elements in X_a should be sorted ascending or descending by their degree. For our example we choose a descending order, which results in an increasing precise cover.

In addition we need to build a second set R_b as temporary storage for the documents and a sorted list of lists $Z = \{Z_1, Z_2, \ldots, Z_k\}$, with the covers Z_i of R_a for the output. The algorithm in pseudocode can be found in Algorithm 2. In every execution of the while loop in line 7 a new sublist $Z_i \subset Z$ is created (see line 13). All of them are complete covers of all documents in R_a, where Z_0 may contain just one element x_i with $N(x_i) = R_a$ and the last Z_m may contain just all identities, that means x_i with a single neighbor $N(x_i) = d_i$. There are many options to modify the algorithm for special use cases. Choosing the ascending order for X_a and the minimum in line 9, which is same as in the other case just means the first $x_j \in X_a$, will mostly give different results.

If after the last run of the loop X_a is empty, but there are still documents in R_a, we receive an incomplete cover Z_k. To avoid that we add the ID's for the last documents in R_a (in descending order) to Z_k, or create and add an all covering x_∞ (for descending order) (Fig. 4).

Algorithm 2 HIERARCHICAL BIPARTITE COVER-DESCRIPTION

Require: Documents $\{d_1, \ldots, d_n\} \subset \mathbb{D}$ and descriptive elements $f(d_i) = \{x_1, \ldots, x_m\} \subset \mathbb{X}$, R_a with all d_i and empty set R_b, sorted list X with all x_i and empty list Z, $G = (R_a \cup X_a, E)$ with $(d_i, x_j) \in E$ if $d_i \in l(x_j)$, order: descending or ascending, maximum iterations maxdeep
Ensure: List of covers Z of $R_a = \{d_1, \ldots, d_n\}$ with elements in $\mathbb{X}$.

```
    for every x_i, x_j ∈ X do
2:    if N(x_i) = N(x_j) then
        x_i = {x_i OR x_j}, remove x_j
4:    end if
    end for
6: k←0
    while |X| > 0 AND k ≤maxdeep do
8:    for every d ∈ R_a do
        choose x_j ∈ N(d_i) with max|N(x_j)| (or min at ascending)
10:       for every d ∈ N(x_j) do
          R_b ← d, from R_a.remove(d)
12:       end for
        move x_j to Z_k
14:     end for
      R_a = R_b, R_b = ∅, k = k + 1
16: end while
    if R_a ≠ ∅ then
18:     if (order = ascending): add x_∞ to last Z_k
      if (order = descending): add f(d_i) for all d_i ∈ R_a to last Z_k
20: end if
    return Z = {Z_1 AND ...AND Z_k}
```

3.4 *Spanning Tree Approach*

Given a set of documents $\mathbb{D}$, a set of context data $\mathbb{X}$ and the document description graph $G = (\mathbb{D} \cup \mathbb{X}, E)$. We can define $\forall x_i \in \mathbb{X}\ D_i = N(x_i)$ as the cover set of x_i in $\mathbb{D}$. We set $D = \{D_1, \ldots, D_n\}$.

A solution of the MDC problem for $R \subset \mathbb{D}$ is a minimum cover $C \subseteq D$ of R so that $C \setminus R$ is minimal.

We can now construct a hierarchical tree using the logical operators *and* and *or* in $\mathbb{X}$. We will do this by considering a directed graph $G' = (V, E)$ with nodes $V = \mathbb{X}$. We add weighted edges between two nodes x_i, x_j if $N_G(x_j) \subset N_G(x_i)$. The weight is set to $w(x_i, x_j) = |N_G(x_i)| - |N_G(x_j)|$. If we add a meta node x_0 that is connected to all nodes that have no nodes adjacent to them, which means to all nodes x with $\delta_G^-(x) = 0$, we can search for minimum spanning trees, see Fig. 5.

Finding the spanning tree(s) in this graph G' can be done using breadth-first search (BFS) or depth-first search (DFS) in $O(|V| + |E|)$ time. Finding the minimum spanning tree can also be done using this approach since the edges are sorted according to their weight. This a technical assumption and we will have different findings on different definitions of $\mathbb{X}$. Finding minimum spanning trees is in general $\mathcal{NP}$-complete, see [21]. See Algorithm 3 for pseudocode.

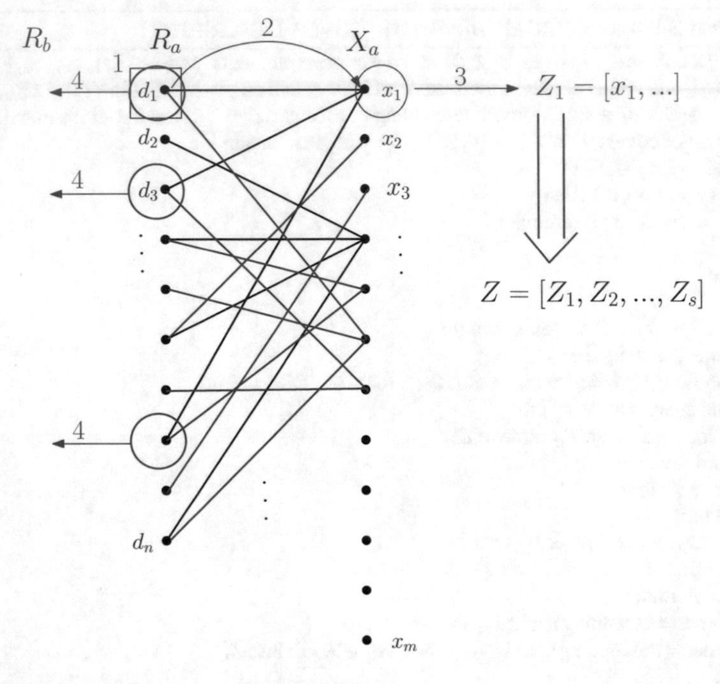

Fig. 4 Illustration of the bipartite graph algorithm

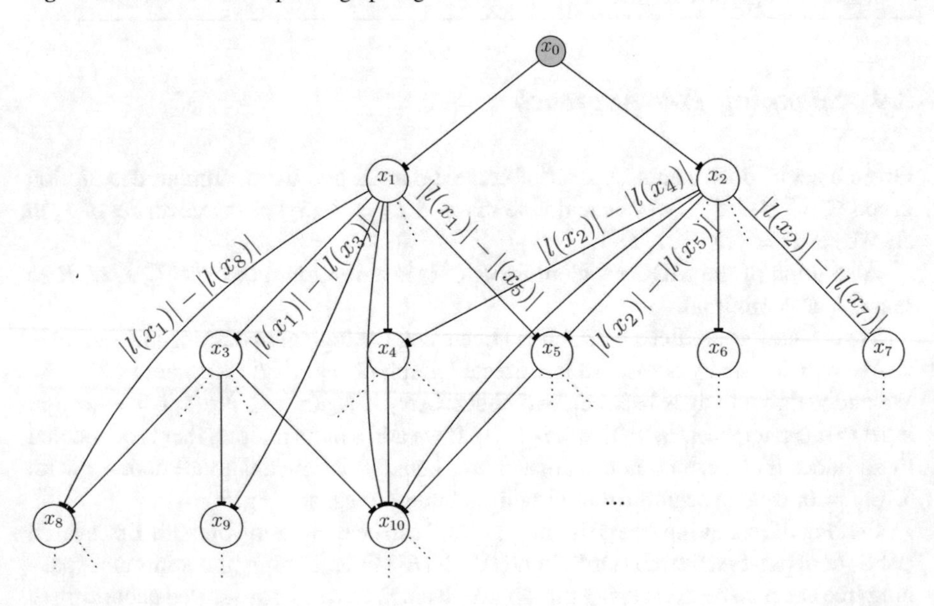

Fig. 5 Illustration of set representative in the graph $G' = (V, E)$ and weight $w(x_i, x_j)$ after adding the meta node x_0, with $l(x_i) := |N_G(x_i)|$. Not all edges and nodes have been added

Algorithm 3 TREE-DESCRIPTION

Require: Documents $d_1, \ldots, d_n \subset \mathbb{D}$ and descriptive elements $f(d_i) = \{x_1, \ldots, x_m\} \subset \mathbb{X}$
Ensure: A spanning tree S describing $R = \{d_1, \ldots, d_n\}$ with elements in $\mathbb{X}$.
1: build list $x_i : l(x_i)$ with $i \in \{1, \ldots, m\}$ and $l(x_i) = q(x_i)$
2: build $G = (X, E)$ with $X = \{x_i, \ldots, x_m\}$ and $(x_i, x_j) \in E$ iff $l(x_j) \subset l(x_i)$ and weight $w(x_i, x_j) = |l(x_i)| - |l(x_j)|$
3: $m = \max_{x \in X} l(x)$
4: $X = X \cup x_o$
5: **for** every $x \in X$ with $l(x) = m$ **do**
6: add edge (x_0, x)
7: **end for**
8: Calculate Minimum Spanning Tree S in G
9: **return** S

As we can see, even this simple approach needs a complex heuristic. Although finding minimum spanning trees is usually in $\mathcal{FP}$, we can construct more complex examples that are $\mathcal{NP}$-complete. It would be very beneficial to find problems that are in $\mathcal{P}$.

4 Non-hierarchical Approaches

4.1 Problem Description

Looking for non-hierarchical approaches we want to find a minimum cover $C \subset D$ without step by step optimization by connecting partial results with logical AND. We here present two ways to do this, first by using an integer linear program and second by using a small modification of the bipartite graph algorithm.

4.2 An Integer Linear Program Approach

Numerous ILP-formulations for the set-cover problem can be found in literature, for example [22] or [23]. To meet Definition 2.6 of MDC we need to adjust the formulation.

Given a set of documents $\mathbb{D}$, a subset $R \subset \mathbb{D}$, a set of context data $f(R) = X \subset \mathbb{X}$ and the document description graph $G = (\mathbb{D} \cup \mathbb{X}, E)$. We can define $\forall x_i \in \mathbb{X}\ D_i = N(x_i)$ as the cover set of x_i in $\mathbb{D}$. We set $D = \{D_1, \ldots, D_n\}$ and $e(D_i) = D_i \setminus R$ as the error of the description term x_i.

A solution of the MDC problem for $R \subset \mathbb{D}$ is a minimum cover $C \subseteq D$ of R so that $C \setminus R$ is minimal.

$$
\begin{array}{ll}
\min & \sum_{i=1}^{n} x_i + \sum_{i=1}^{n} x_i e(X_i) \\
\text{subject to} & \sum_{i:v\in X_i} x_i \geq 1, \ \forall v \in R \\
& x_i \geq 1 \ \forall i = 1, \ldots, n \\
& x_i \in \mathbb{Z} \ \forall i = 1, \ldots, n
\end{array}
\tag{2}
$$

Here the vector x gives a set $Z \subset X$ which gives a minimum cover $q(Z) = C \subset D$ of R so that $C \setminus R$ is minimal.

The weighted MDC problem was introduced in Definition 2.7. Given a weight function $w : \mathbb{X} \to \mathbb{R}$ that defines a weight for every element in $\mathbb{X}$ the ILP (2) changes as follows:

$$
\begin{array}{ll}
\min & \sum_{i=1}^{n} w(x_i) + \sum_{i=1}^{n} x_i e(X_i) \\
\text{subject to} & \sum_{i:v\in X_i} x_i \geq 1 \ \forall v \in R \\
& x_i \geq 1 \ \forall i = 1, \ldots, n \\
& x_i \in \mathbb{Z} \ \forall i = 1, \ldots, n
\end{array}
\tag{3}
$$

A solution of the MDC problem for $R \subset \mathbb{D}$ is a minimum cover $C \subseteq D$ of R, i.e. $\sum_{c\in C} w(c)$ is minimal, so that $C \setminus D$ is minimal.

4.3 *Dynamic Programming and Bipartite Graph Heuristic*

We can use Algorithm 2 to construct a non-hierarchical solution. This algorithm has already been used to computed k covers of R_a, which can be used to find a cover with minimal error $Z = \min_{e(x_i)} Z_i$, that means for $q(Z) = C$ $C \setminus R$ is minimal. The pre-sorting of the context data list X results in covers of ascending cardinality, so the number of iterations k may be a limit for maximum cardinality. The pre-sorting can be removed, which results in more balanced and random covers, whereof one with minimum error can be chosen.

5 Experimental Results

We tested our novel approach within two scenarios. First of all, using an artificial random instances with $|\mathbb{D}| = 150$ documents and a given subset R with 20 example documents. We created instances with a fixed number of 80 or 40 normal distributed keywords which had a significant impact on the output. In addition we used N iterations, which lead to a different precision. The second scenario is a real-world example using set R of 10 random documents out of a human curated topic. We tested against complete PubMed Database using SCAIView. Thus $|\mathbb{D}| \approx 29{,}000{,}000$.

Within the random instances we were unable to describe a single document by its random keywords. This approach usually returned more than 100 documents. The reason for this rather contradictory result is still not entirely clear, but the normal distribution of keywords may be responsible for this result. The algorithms Tree-

Algorithm 4 BIPARTITE COVER-DESCRIPTION

Require: Documents $\{d_1, \ldots, d_n\} \subset \mathbb{D}$ and descriptive elements $f(d_i) = \{x_1, \ldots, x_m\} \subset \mathbb{X}$, R_a with all d_i and empty set R_b, sorted list X with all x_i and empty list C, $G = (R_a \cup X_a, E)$ with $(d_i, x_j) \in E$ if $d_i \in N(x_j)$, maximum iterations maxdeep
Ensure: A minimum covers Z of $R_a = \{d_1, \ldots, d_n\}$ with elements in $\mathbb{X}$.

```
    for every x_i, x_j ∈ X do
2:    if N(x_i) = N(x_j) then
        x_i = {x_i OR x_j}, remove x_j
4:    end if
    end for
6: k←0
    while |X| > 0 AND k ≤maxdeep do
8:    for every d ∈ R_a do
        choose x_j ∈ N(d_i) with max|N(x_j)|
10:       for every d ∈ N(x_j) do
            R_b ← d, from R_a.remove(d)
12:       end for
        move x_j to Z_k
14:     end for
      R_a = R_b, R_b = ∅, k = k + 1
16: end while
    if R_a ≠ ∅ then
18:     add x_∞ to last Z_k
    end if
20: return Z = min_{i∈{1,...,k}} Z_i,
```

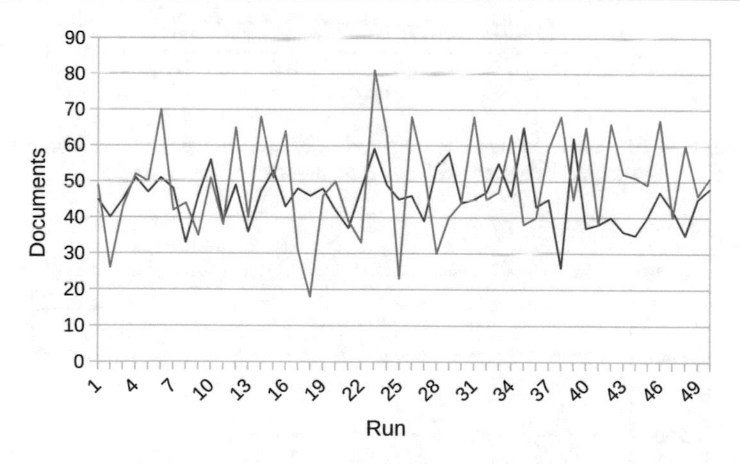

Fig. 6 Output of 50 random example runs and the number of retrieved documents in the artificial random scenario for algorithms Tree-Description (green) and Hierarchical Bipartite Cover-Description (red). The total number of documents was 150, and the document subset contains 20 documents. The number of keywords was 40. The number of iterations is $N = 4$

Description and Hierarchical Bipartite Cover-Description performed quite well, see Fig. 6. In general, we found Hierarchical Bipartite Cover-Description to work better and faster.

Changing to the real-world scenario the situation changes significantly. Given a set of 10 documents, Hierarchical Bipartite Cover-Description usually returned more than 6,000,000 documents, Tree-Description more than 5,000,000 before reaching the search-query length limitations. Vice versa we found, that the combination of keywords described a single document very well—even within nearly 3 million documents in $\mathbb{D}$. The keywords using MeSH-terms in PubMed are manually curated and seem not to be normally distributed.

The output of Keyword-Cover for 10 random examples with $|R| = 10$ is presented in Figs. 7 and 8. The precision was iterated from 0.9 to 0.4. The output scales very well

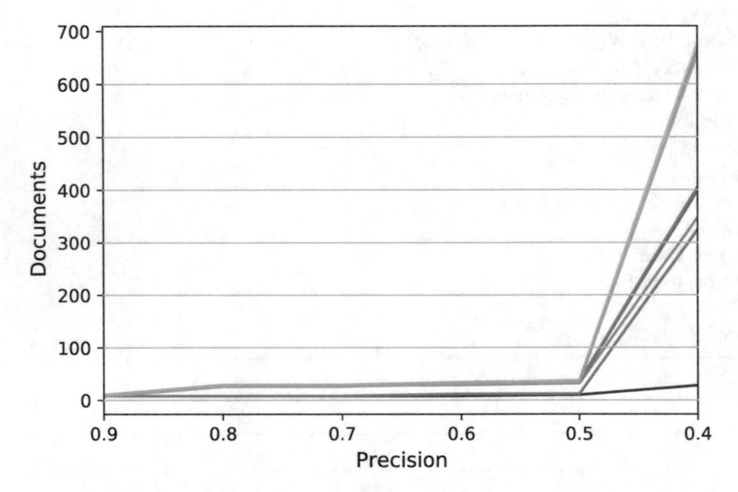

Fig. 7 Output of 10 random example runs with $|R| = 10$ on PubMed. The precision was iterated from 0.9 to 0.4. The output scales very well and is quite stable till precision 0.5

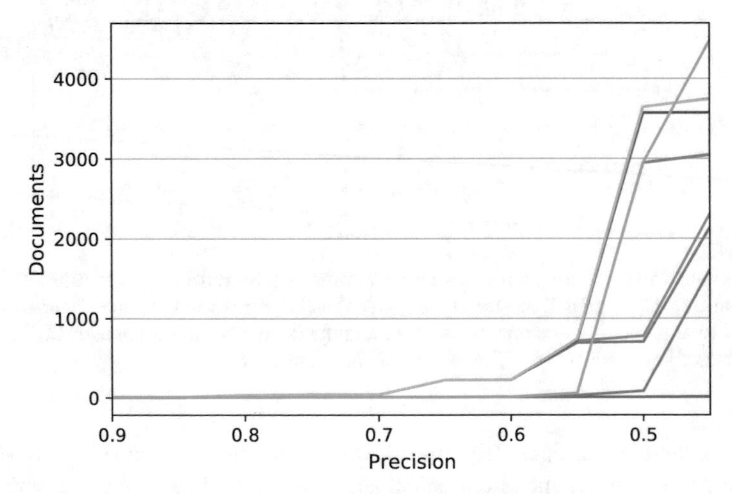

Fig. 8 Output of 10 random example runs with $|R| = 10$ on PubMed. The precision was iterated from 0.9 to 0.4. The output scales very well and is quite stable till precision 0.5

and is quite stable till precision 0.5 where we found between 12 and 36 documents. For precision 0.4 we found 28 till 676 documents.

We can see, that we have found a novel solution for search query finding on literature that performs quite well on real-world data. Our work clearly has some limitations. It is not clear, why the proposed algorithms perform significantly different in both scenarios. Despite this we believe our work could be the basis for solving the SQF and TD. Further work needs to be performed to the distribution of descriptive elements to documents to establish whether they can be used to generate search queries and topic descriptions that are significant enough.

6 Conclusions

We presented a novel formulation of both search query and topic finding problems as Minimum Set-Cover Problems. We proposed a weighted and unweighted version of the Minimum Document-Cover Problem as well as a hierarchical version using both AND as well as OR and the non-hierarchical version only using and.

With this we get a solution that uses on the one hand as much descriptive elements as possible to get as less documents in $\mathbb{D}$ but not in R.

The search queries are not human readable. For example the tree-approach returns queries in the form MeSH_Terms: D000818" AND ("MeSH_Terms: D051381" OR "MeSH_Terms: D009538" OR "MeSH_Terms: D017207" OR "MeSH_Terms: Q000494" OR "MeSH_Terms: D006624" OR "MeSH_Terms: D011978" OR "MeSH_Terms: D000109" OR "MeSH_Terms: D008297" OR "MeSH_Terms: Q000187" OR "MeSH_Terms: Q000502" OR "MeSH_Terms: Q000378" OR "MeSH_Terms: D008464" OR "MeSH_Terms: Q000187" OR "MeSH_Terms: Q000187" OR .... This can be easily translated into something human-readable. But still it is a good probability that further research has to be done on how to shorten this to be both precise as well as significant.

In general this is both: a correct solution of clustering labeling of R on $\mathbb{X}$ obtained by f as well as a possible solution of a search query so that $q(Z) = R$. It is not necessary an optimal solution of SQF or CLF problem, since reordering the keywords may result in better solutions.

The bipartite graph algorithms can be modified for many different use cases. All hierarchical algorithms can also be modified by adding weights. As described, there are many possible variations like sorting the context data list by minimum or maximum degree. The number of iterations k also has a big impact on the result. Another possible optimization is the pre-sorting by weighting the x_i with maximum $|N(x_i)|$ and minimal $D \setminus R$.

This paper has underlined the importance of finding the computational core of NLP problems. We have managed to find a Minimum Set-Cover reformulation of SQF and TF which lead to an accurate solving of both on real-world data. The current study was unable to reproduce this success on random input data. Thus it is recommend that further research should be undertaken to examine the impact of keyword (or descriptive elements) distributions on documents. Nevertheless these results have been very encouraging to integrate this feature in SCAIView and to do further research on the optimization and extension of this heuristic.

References

1. Dörpinghaus, J., Darms, J. Jacobs, M.: What was the question? a systematization of information retrieval and NLP problems. In: 2018 Federated Conference on Computer Science and Information Systems (FedCSIS). IEEE (2018)
2. Coordinators, N.R.: Database resources of the national center for biotechnology information. Nucl. Acids Res. **45**(Database issue), D12 (2017)
3. Hodapp, S., Madan, S., Fluck, J., Zimmermann, M.: Integration of UIMA text mining components into an event-based asynchronous microservice architecture. In: Proceedings of the LREC 2016 Workshop "Cross-Platform Text Mining and Natural Language Processing Interoperability", p. 72 (2016) [Online]. Available: https://pdfs.semanticscholar.org/3f43/ea96912065699199dcf1378fbbfd00929119.pdf
4. Emon, M.A.E.K., Karki, R., Younesi, E., Hofmann-Apitius, M., et al.: Using drugs as molecular probes: a computational chemical biology approach in neurodegenerative diseases. J. Alzheimer's Dis. **56**(2), 677–686 (2017)
5. Iyappan, A., Younesi, E., Redolfi, A., Vrooman, H., Khanna, S., Frisoni, G.B., Hofmann-Apitius, M.: Neuroimaging feature terminology: a controlled terminology for the annotation of brain imaging features. J. Alzheimer's Dis. **59**(4), 1153–1169 (2017)
6. Dörpinghaus, J., Schaaf, S., Fluck, J., Jacobs, M.: Document clustering using a graph covering with pseudostable sets. In: 2017 Federated Conference on Computer Science and Information Systems (FedCSIS), pp. 329–338. IEEE (2017)
7. Madan, S., Hodapp, S., Senger, P., Ansari, S., Szostak, J., Hoeng, J., Peitsch, M., Fluck, J.: The BEL information extraction workflow (BELIEF): evaluation in the BioCreative V BEL and IAT track. In: Database, vol. 2016, p. baw136, Oct 2016 [Online]. Available: http://database.oxfordjournals.org/lookup/doi/10.1093/database/baw136
8. Szostak, J., Ansari, S., Madan, S., Fluck, J., Talikka, M., Iskandar, A., De León, H., Hofmann-Apitius, M., Peitsch, M.C., Hoeng, J.: Construction of biological networks from unstructured information based on a semi-automated curation workflow. Database J. Biol. Databases Curation **2015** (2015)
9. Suryanarayana, D., Hussain, S.M., Kanakam, P., Gupta, S.: Natural language query to formal syntax for querying semantic web documents. In: Progress in Advanced Computing and Intelligent Engineering, pp. 631–637. Springer (2018)
10. Melo, D., Rodrigues, I.P., Nogueira, V.B.: Semantic web search through natural language dialogues. In: Innovations, Developments, and Applications of Semantic Web and Information Systems. IGI Global, pp. 329–349 (2018)
11. Lin, J., Wilbur, W.J.: Pubmed related articles: a probabilistic topic-based model for content similarity. BMC Bioinform. **8**(1), 423 (2007)
12. Newman, D., Karimi, S., Cavedon, L.: Using topic models to interpret medline's medical subject headings. In: Australasian Joint Conference on Artificial Intelligence, pp. 270–279. Springer (2009)
13. Trieschnigg, D., Pezik, P., Lee, V., De Jong, F., Kraaij, W., Rebholz-Schuhmann, D.: Mesh up: effective mesh text classification for improved document retrieval. Bioinformatics **25**(11), 1412–1418 (2009)

14. Lu, Z., Wilbur, W.J., McEntyre, J.R., Iskhakov, A., Szilagyi, L.: Finding query suggestions for pubmed. In: AMIA Annual Symposium Proceedings, vol. 2009, p. 396. American Medical Informatics Association (2009)
15. Bertossi, A.A.: Dominating sets for split and bipartite graphs. Inf. Process. Lett. **19**(1), 37–40 (1984)
16. Yannakakis, M., Gavril, F.: Edge dominating sets in graphs. SIAM J. Appl. Math. **38**(3), 364–372 (1980)
17. Korte, B., Vygen, J., Korte, B., Vygen, J.: Combinatorial Optimization, vol. 2. Springer (2012)
18. Garey, M.R., Johnson, D.S.: Computers and Intractability, vol. 29. W.H. Freeman, New York (2002)
19. Karp, R.M.: Reducibility among combinatorial problems. In: Complexity of Computer Computations, pp. 85–103. Springer (1972)
20. Hagen, M., Michel, M., Stein, B.: What was the query? generating queries for document sets with applications in cluster labeling. In: International Conference on Applications of Natural Language to Information Systems, pp. 124–133. Springer (2015)
21. Camerini, P., Galbiati, G., Maffioli, F.: Complexity of spanning tree problems: part I. Eur. J. Oper. Res. **5**(5), 346–352 (1980) [Online]. Available: http://www.sciencedirect.com/science/article/pii/0377221780901642
22. Balas, E., Padberg, M.W.: On the set-covering problem. Oper. Res. **20**(6), 1152–1161 (1972)
23. Vazirani, V.V.: Approximation Algorithms. Springer Science & Business Media (2013)

Evaluation of Optimal Charging Station Location for Electric Vehicles: An Italian Case-Study

Edoardo Fadda, Daniele Manerba, Gianpiero Cabodi, Paolo Camurati, and Roberto Tadei

Abstract Electric vehicles are accelerating the world transition to sustainable energy. Nevertheless, the lack of a proper charging station infrastructure in many real implementations still represents an obstacle for the spread of such a technology. In this paper, we present a real-case application of optimization techniques in order to solve the location problem of electric charging stations in the district of Biella, Italy. The plan is composed by several progressive installations and the decision makers pursue several objectives that might conflict each other. For this reason, we present an innovative framework based on the comparison of several ad-hoc Key Performance Indicators (KPIs) for evaluating many different location aspects.

Keywords Electric vehicles · Charging stations · Optimal location · KPIs

This work has been supported by Ener.bit S.r.l. (Biella, Italy) under the research projects "Studio di fattibilità per la realizzazione di una rete per la mobilità elettrica nella provincia di Biella" and "Analisi per la realizzazione di una rete per la mobilità elettrica nella provincia di Biella". The authors want to acknowledge Prof. Guido Perboli, Politecnico di Torino, for his contribution to derive the demand analysis presented in Sect. 3.

E. Fadda (✉) · G. Cabodi · P. Camurati · R. Tadei
Department of Control and Computer Engineering, Politecnico di Torino, Turin, Italy
e-mail: edoardo.fadda@polito.it

G. Cabodi
e-mail: gianpiero.cabodi@polito.it

P. Camurati
e-mail: paolo.camurati@polito.it

R. Tadei
e-mail: roberto.tadei@polito.it

D. Manerba
Department of Information Engineering, University of Brescia, Brescia, Italy
e-mail: daniele.manerba@unibs.it

S. Fidanova (ed.), *Recent Advances in Computational Optimization*, Studies in Computational Intelligence 920,
https://doi.org/10.1007/978-3-030-58884-7_4

1 Introduction

With the increasing pressure on the environment and resource shortage, energy saving has become a global concern. However, this issue is particularly critical in the field of transportation for freight and people. In fact, it has been estimated that motorized vehicles are responsible for 40% of carbon dioxide emissions and 70% of other greenhouse gas emissions in urban areas [12]. This has led to the consideration of alternatives to the current mobility and, due to the technology development, electric vehicles (EVs) have become a clean and sustainable alternative to traditional fuel ones. However, one of the barriers that still limits the desirable expansion of EVs industry is the lack of a proper infrastructure for re-charging the vehicles or, more in general, of a structured guideline for the administrations to decide where to locate the available charging stations so to optimize the quality of the service.

In this context, the company *Ener.bit S.r.l.*[1] and the *Dipartimento di Automatica e Informatica (Control and Computer Engineering Department)* of the Politecnico di Torino (Polytechnic University of Turin, Italy) have recently developed a project for the sustainability of electric mobility in the district of Biella, Piedmont (Italy). The project goal was to plan the type, number, and location of the charging stations over a horizon of about 10 years (2019–2030). According to PNire,[2] i.e. the Italian infrastructural plan for EV charge, the possible infrastructures that can be build are slow charging (up to 7kW), quick charging (between 7 and 22 kW), fast charging (between 22 and 50 kW), and very fast charging (more than 50 kW) stations. Several strategic areas characterized by different capacity as well as different stopping time (parking areas, shopping centers, railway stations, etc.) have been identified as possible places where to locate the charging stations. It is worthwhile noticing that the number of stations to locate depends on an economical analysis of the decision process over the time horizon, whereas the type of charging stations mainly depends on the features of the selected location and on the analysis of the traffic flow (see [9–11]).

For example, a charging station near working centers can have a slow charging system (because workers are assumed to park their vehicle during the working hours, almost eight), whereas a charging station near shopping centers must be faster (cars must be recharged during the shopping time, up to two hours). Therefore, the actual operational problem faced by our project team was to identify an optimal location of the different charging stations in the various municipalities of the district.

[1] Official website: http://www.enerbit.it/, last accessed: 2020-01-29.

[2] http://www.governo.it/sites/governo.it/files/PNire.pdf, last accessed: 2020-01-29.

The good results obtained in this project have fostered a wider and deeper analysis of the problem of locating charging stations for electric vehicles (see the first results in [7]). In fact, location problems still attract great attention from the research community (see, e.g., [3, 16]).

Usually, the real decision maker despite defining a single objective (such as cost minimization or gain maximization) are interested in several aspects of the solution. Thus, it is not rare to evaluate the solution with respect to several criteria not explicitly considered in the objective function (see, e.g., [6, 8]). In particular, location problems may consider several different (and possibly conflicting) objectives, e.g., achieving a level of service proportional to the importance of the location, reducing the worst-case service level, and maximizing the average service level. Considering all those objectives in the same mathematical problem may end up with a huge amount of solutions that can confuse the decision maker instead of providing help. For this reason, our study provides an innovative analysis based on the comparison of several different aspects of a location solution through the use of a battery of Key Performance Indicators (KPIs). Moreover, since charging infrastructures are commonly supposed to be located through several progressive interventions over a defined time-horizon, we also analyze the trend of the provided KPIs over the interventions to generate long-term managerial insights.

The rest of this paper is organized as follows. In Sect. 2, a review of the literature regarding location of electric vehicle is given. In Sect. 3, the case study is presented. Section 4 is devoted to present the location model used in the project. In Sect. 5, we propose and discuss several different KPIs of interest for our application. In Sect. 6, we present the numerical results. Finally, conclusions are drawn in Sect. 7.

2 Literature Review

A great number of applications in the field of electrical vehicles have appeared in the literature, and several aspects have been studied from the point of view of optimization. In particular, the computation of an optimal location of the charging stations seems of fundamental importance. In the following, we will review the most important and recent works related to this problem.

In [12] the authors present a study on the location of electric-vehicle charging stations for the city of Lisbon (Portugal), characterized by a strong concentration of population and employment. This type of area is appropriate for slow charging because vehicles remain parked for several hours within 24 h period. The methodology is based on a maximal covering model to optimize the demand covered within an acceptable level of service and to define the number and capacity of the stations to be installed. They proposed a complex model maximizing demand coverage, distinguishing between night-time and day-time demand.

In [1, 19] the authors develop a complex model that optimally locates the charging stations by considering the travel patterns of individual vehicles. The model is applied to the city of Beijing (China) using vehicle trajectory data of 11,880 taxis over a period of three weeks. They use the taxi fleet as a case-study because public fleets are likely to be early adopters for electric vehicles. Similarly, in [15] the authors consider a bi-level programming model including electric vehicle driving range for finding an optimal location of charging stations. Similar approaches can be found in [17, 23].

Considering a very similar setting, in [25] the authors formulate a multi-period optimization model based on a flow-refueling location model for strategic charging station location planning. They consider that, although it is expected that a sufficient number of charging stations will be eventually constructed, due to various practical reasons they may have to be introduced gradually over time. They simultaneously optimize the problem of where to locate the charging stations and how many chargers should be established in each charging station. By considering both decisions together, the complexity of the model increases, thus the authors propose a genetic algorithm-based method for solving the expanded model. Almost the same approach is followed by [2] for the city of Seattle (U.S.A.). In this paper, the authors consider a p-center problem enriched by the parking capacity problem. Other similar studies can be found in [21, 24].

In [4], the authors propose a Mixed Integer Linear Programming (MILP) model to solve the plug-in hybrid electric vehicles (PHEV) charging infrastructure planning problem for organizations with thousands of people working within a defined geographic location and parking lots well-suited to charging station installations. Finally, [13] proposes a maximum covering model to locate a fixed number of charging stations in central urban areas to maximize the demand covered within a given distance, where the demand of each study area is determined by estimating the number of vehicles in the area.

As the reader can notice, the majority of the works define an ad-hoc optimization model describing some particular feature of the application. The goal of the present paper is to revert that paradigm: consider a standard model and measure the characteristic of the solution with respect to the performance indicator usually considered as goals. This approach, to the author knowledge, has been never considered in previous works.

3 An Italian Case-Study

The use-case considered deals with the location of electric charging stations in the district of Biella (Italy). In particular, the potential locations considered by the company are the 78 municipalities of the district. Thus, the main optimization problem is to decide in which municipalities to locate some charging stations. To do that, we first need to compute the total number of charging stations to be located, which is based on the expected number of electric vehicles.

From 2016 the registration of electric vehicles in the world (including electric cars, plug-in hybrids and electric fuel cells) is increasing, with over 750,000 sales globally. With a market share of 29%, Norway is confirmed as one of the leaders in the electric mobility revolution. It is followed by the Netherlands (6.4%) and Sweden (3.4%) and then by China, France, and the UK (1.5%). Despite promising estimates, there is strong uncertainty about the impact of electric mobility. In Italy, the uncertainty about battery life is the main barrier to the adoption of electric vehicles (35%), followed by the presence of charging stations (34%) and the low presence of fast charging stations (17%). Thus, the policies implemented by the regulator and the correct planning of the charging infrastructure are of fundamental importance for the development of electric vehicles. Against about 2 million vehicles sold in 2016, the penetration of electric vehicles settles at less than 0.1%. Annual sales in Italy settle at 1400 units/year, with a fleet of 5657 cars. In order to estimate the number of electric vehicles in the district of Biella, we assume that the propensity to use electrical technology in the district of Biella is the same as that of the rest of Italy. By crossing this data with the vehicles sold in Piedmont and the absorption by the district of Biella, we obtain an estimate of approximately 35 vehicles sold per year and 270 vehicles in the park circulating at the current date, on a total of approximately 152,000 vehicles. In order to determine the future number of electric vehicles, it is important to estimate the diffusion of the technology. The most used model in the literature is the sigmoidal function in Fig. 1, obtained from the integration of a normal curve.

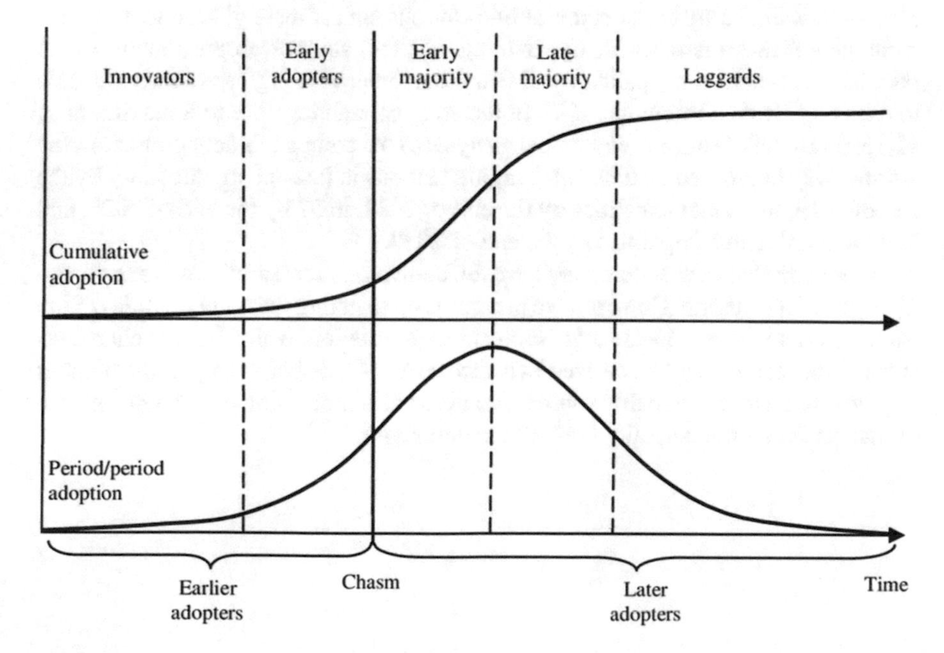

Fig. 1 Representation of the sigmoid function

The curve can be divided into three different phases:

- a first phase, in which the curve grows slowly, in which users are the so-called *innovators*;
- a phase of rapid adoption, in which users are the so-called early adopters and early majority, that is users who require mature technology, but are willing to pay more than others for its access;
- a last phase of reduction of market penetration (due to the saturation of the same), in which users want a service with maximum efficiency and minimum cost (the so called late majority and laggards).

The estimate of the parameters of the diffusion curve starting from historical data therefore makes it possible to estimate the effective size of the market and the adoption factors. A universally recognized approach is that of the so-called Bass diffusion model (see [18]). To this end, we have set up a model for estimating parameters using the R statistical software and the Diffusion package, starting from the sales data of electric vehicles in Italy 2009–2016 [20]. The electric car fleet in the district of Biella in 2030 can therefore be estimated at 56,000 cars, equal to about 30% of the car fleet in circulation at that time and an electricity market share of 13%. The impact of freight transport (no more than 100 expected electric vehicles) is considered limited and then disregarded.

Since the key factor for the adoption of electric mobility is the availability of fast charging stations (and some ultra-fast charging ones), making the hypothesis of an average charge of 2 h and a use for about 16 h a day of the recharging and that only 70% of the circulating car fleet use public stations, an estimate of around 4900 fast recharging stations is reached. By dividing all these stations proportionally to the population in each municipality, we obtain the number of charging stations to locate in each city (from a minimum of 72 in the smallest municipality to a maximum of 41,139 in Biella). From an economical analysis of the company's economic flow, the optimal way to proceed is to install charging stations in just one municipality by the end of 2019, in 10 municipalities by the end of 2022, in 37 by the end of 2025, and in all remaining municipalities by the end of 2030.

We remark that each station may have different size, number of plugs, and capacity in terms of charging. However, we just focus on selecting the municipalities of the Biella district where to locate at least one charging station, while the real characteristics of the stations will be derived in a successive phase. For example, the number of plugs for each municipality can be calculated as a proportion to the demand rate of that particular municipality (and its surroundings).

4 Mathematical Models for Optimal Location

In this section, we describe the classical p-median, p-center, and p-centdian models to find the optimal location for the charging stations since, despite of their simplicity, they well describe the main goal of the company. Furthermore, since these models are easy to solve in practice, it is possible to compute several solutions with different inputs in a short amount of time.

In the rest of this section and throughout the whole paper we use the following notation:

- $G = (N, E)$: complete undirected graph with a set of nodes N representing possible locations for the charging stations and a set of edges $E = \{(i, j)|i, j \in N, i \leq j\}$;
- d_{ij}: distance between node i and node $j \in N$ (note that distance d_{ii} may be non-null since it represents the internal distance to travel within municipality $i \in N$);
- Q_i: service demand in node $i \in N$;
- $h_i = Q_i / \sum_{j \in N} Q_j$: demand rate of node $i \in N$;
- p: predefined number of stations to locate, with $p \leq |N|$;
- $\bar{d}$: coverage radius, i.e. the threshold distance to discriminate the covering. It represents, e.g., the maximum distance that an EV can travel (due to the battery capacity) or that a user is willing to drive to reach a charging station;
- $C_i = \{j \in N, d_{ij} \leq \bar{d}\}$: covering set of $i \in N$, i.e. the set of all stations nearer than $\bar{d}$ from node i.

4.1 p-Median

The *p-median* problem is to find p nodes of the network in which to locate a charging station so to minimize the weighted average distance between the located stations and the demand nodes. It can be stated as

$$\min \sum_{i \in N} h_i \sum_{j \in N | (i,j) \in E} d_{ij} x_{ij} \tag{1}$$

subject to

$$\sum_{j \in N | (i,j) \in E} x_{ij} = 1 \quad \forall i \in N \tag{2}$$

$$\sum_{j \in N} y_j = p \tag{3}$$

$$\sum_{i \in N | (i,j) \in E} x_{ij} \leq |N| y_j \quad \forall j \in N \tag{4}$$

$$y_j \in \{0, 1\}, \quad \forall j \in N \tag{5}$$

$$x_{ij} \in \{0, 1\}, \quad \forall (i, j) \in E \tag{6}$$

where x_{ij} is a binary variable for each edge $(i, j) \in E$ that takes value 1 iff the demand of node $i \in N$ is served by a charging station located in $j \in N$. The objective function (1) consists of minimizing the average distance traveled by the total demand flow towards charging stations. Constraints (2) ensure that each demand node is served by exactly one station. Constraint (3) ensures to locate exactly p stations. Logical constraints (4) ensure to locate a station in j (i.e., $y_j = 1$) if it is assigned to serve at least one demand node (i.e., $\sum_{i \in N | (i,j) \in E} x_{ij} > 0$). Finally, (5) and (6) state binary conditions on the variables.

4.2 p-Center

The *p-center* problem is to find p nodes where to locate charging stations so to minimize the maximum distance between a demand node and its closest station. In the proposed version of the problem, sometimes called *vertex restricted p-center* problem, the stations can be located only in the nodes of the graph. Obviously, the problem is focused on the worst case in terms of distance and can be stated as

$$\min M \tag{7}$$

subject to

$$M \geq \sum_{j \in N | (i,j) \in E} h_i d_{ij} x_{ij} \quad \forall i \in N \tag{8}$$

and the already presented constraints (2)–(6). The objective function (7) aims at minimizing an auxiliary variable M that, according to constraints (8), will take the maximum value of the expression $\sum_{j \in N} h_i d_{ij} x_{ij}$ over all the nodes $i \in N$.

4.3 p-Centdian

The *p-centdian* problem is to find p nodes where to locate charging stations so to minimize a linear combination of the objective functions of the *p-median* and *p-center* problems. Thus, the p-centdian has characteristics in between the the p-center and p-median. The formulation is as follows

$$\min \lambda M + (1 - \lambda) \sum_{i \in N} h_i \sum_{j \in N | (i,j) \in E} d_{ij} x_{ij} \tag{9}$$

subject to

$$M \geq \sum_{j \in N | (i,j) \in E} h_i d_{ij} x_{ij} \quad \forall i \in N \tag{10}$$

and the already presented constraints (2)–(6). Through the parameter λ, with $0 \leq \lambda \leq 1$, it is possible to define the relative importance of one objective with respect to the other one. In this work, we set the parameter λ dynamically by using the optima of the p-center and p-median subproblems and calibrating their combination in order to have the two terms with the same magnitude.

5 Key Performance Indicators

In this section, we define the set of KPIs that were used in the project in order to measure the performance of the solution provided by the model. For simplicity, we define $\mathcal{L}_i = \{j \in \mathcal{C}_i \mid y_j = 1\}$ as the set of nodes where a charging station has been located that covers demand node i, and $\mathcal{C} = \{i \in N \mid \exists j \in \mathcal{C}_i \text{ such that } y_j = 1\}$ as the set of demand nodes covered by at least one charging station.

The proposed KPIs consider topological, coverage, and accessibility measures. They are summarized in Table 1 and detailed explained in the following:

- Worst-case distance: Eq. (11) represents the maximum distance between a demand node and its closest charging station.
- Weight of the worst-case distance: Eq. (12) represents the demand rate that is affected by the worst-case scenario in terms of distance.
- Best-case distance: Eq. (13) represents the minimum distance between a demand node and its closest charging station.
- Weight of the best-case distance: Eq. (14) represents the demand rate that is affected by the best-case scenario in terms of distance.
- Average distance: Eq. (15) represents the average distance between a demand node and its closest charging station.
- Weighted average distance: Eq. (16) represents the average distance in which each node is weighted by its demand rate.
- Dispersion: Eq. (17) represents the sum of the distances between all the located stations. It is a measure of homogeneity of the service from a purely topological point of view.

Table 1 KPIs definition

Description	Name	Formula	
Worst-case distance	D_{max}	$\max_{i \in N} \min_{j \in \mathcal{L}} d_{ij}$	(11)
Weight of the worst-case distance	D^h_{max}	h_i such that $\arg\max_{i \in N} \min_{j \in \mathcal{L}} d_{ij}$	(12)
Best-case distance	D_{min}	$min_{i \in N} \min_{j \in \mathcal{L}} d_{ij}$	(13)
Weight of the best-case distance	D^h_{min}	h_i such that $\arg\min_{i \in N} \min_{j \in \mathcal{L}} d_{ij}$	(14)
Average distance	D_{avg}	$\frac{1}{\lvert N \rvert} \sum_{i \in N} \min_{j \in \mathcal{L}} d_{ij}$	(15)
Weighted average distance	D^h_{avg}	$\frac{1}{\lvert N \rvert} \sum_{i \in N} \min_{j \in \mathcal{L}} h_i d_{ij}$	(16)
Dispersion	$Disp$	$\sum_{i \in \mathcal{L}} \sum_{j \in \mathcal{L}} d_{ij}$	(17)
Accessibility	Acc	$\sum_{i \in N} h_i A_i, \text{ with } A_i := \sum_{j \in \mathcal{L}} e^{-\beta d_{ij}}$	(18)
Coverage	C	$\lvert \mathcal{C} \rvert / \lvert N \rvert$	(19)
Weighted coverage	C^h	$\sum_{i \in \mathcal{C}} h_i$	(20)
Weight of redundant coverage	RC^h	$\sum_{i \in N} \sum_{j \in \mathcal{L}_i} h_i$	(21)
Worst-case coverage	C_{min}	$\min_{i \in N} \lvert \mathcal{L}_i \rvert$	(22)
Weight of the worst-case coverage	C^h_{min}	h_i such that $\arg\min_{i \in N} \lvert \mathcal{L}_i \rvert$	(23)
Best-case coverage	C_{max}	$\max_{i \in N} \lvert \mathcal{L}_i \rvert$	(24)
Weight of the best-case coverage	C^h_{max}	h_i such that $\arg\min_{i \in N} \lvert \mathcal{L}_i \rvert$	(25)
Average coverage	C_{avg}	$\frac{1}{N} \sum_{i \in N} \lvert \mathcal{L}_i \rvert$	(26)
Weighted average coverage	C^h_{avg}	$\frac{1}{N} \sum_{i \in N} h_i \lvert \mathcal{L}_i \rvert$	(27)

- Accessibility: Eq. (18) is the total accessibility of the charging service, where

$$A_i := \sum_{j \in \mathcal{L}} e^{-\beta d_{ij}} \tag{28}$$

is the accessibility of a facility in the sense of [14]. The parameter $\beta > 0$ must be calibrated and represents the dispersion of the alternatives in the choice process (the calibration has been performed according to [5, 22]).
- Coverage: Eq. (19) represents, in percentage, the number of covered locations with respect to the total.
- Weighted coverage: Eq. (20) represents, in percentage, the demand rate of the covered locations with respect to the total demand (we remark that, by definition, $\sum_{i \in N} h_i = 1$).
- Weight of the redundant coverage: Eq. (21) represents, in percentage, the demand rate of the covered locations multiplied by the times that such locations are covered. This indicator measures the weighted redundancy of the coverage.
- Worst-case coverage: Eq. (22) represents the minimum number of charging stations covering a demand node.
- Weight of the worst-case coverage: Eq. (23) represents the demand rate that is affected by the worst-case scenario in terms of coverage.
- Best-case coverage: Eq. (24) represents the maximum number of charging stations covering a demand node.
- Weight of the best-case coverage: Eq. (25) represents the demand rate that is affected by the best-case scenario in terms of coverage.
- Average coverage: Eq. (26) represents the average number of charging stations covering a demand node.
- Weighted average coverage: Eq. (27) represents the average coverage in which each node is weighted by its demand rate.

6 Numerical Experiments

The instances of the Biella problem are generated according to real data. In particular, the matrix of distance d_{ij} considers the time, in minutes, to travel from i to j. The diagonal elements (i.e., $d_{ii}, \forall i \in N$) are estimated according to the geographical extension of the city. The estimations are considered from the Istat (Istituto Nazionale di Statistica) website.[3]

Among the three proposed and described in Sect. 4, the model chosen by the company for the solution of the problem is the p-centdian. This choice is due to its flexibility with respect to the company goal and to its global performance in terms ok KPIs (which has appeared to be better than the one of the p-center and p-median

[3]http://www.istat.it/storage/cartografia/matrici_distanze/Piemonte.zip.

models in some preliminary experiments). The p-centdian model, accurately instantiated with the data deriving from the Biella district case study, can be easily solved by exact algorithms as the branch-and-cut implemented in the available commercial and academic solvers. In our particular case, we used the GUROBI solver v.8.1.0. The resolution was performed on a common PC (Intel Core i7-5500U CPU@2.40 GHz with 8 GB RAM) and took on average 12 s. Notice how the resolution efficiency obtained allows to possibly perform a large number of experiments with different input data, thus refining the analysis.

The solutions for the different time thresholds studied, obtained using the p-centdian model, are the following (clearly, at each intervention, the locations chosen in the previous steps are forced to remain in the solution):

- one municipality ($p = 1$) by the end of 2019: the only municipality chosen is Biella, the chief town (see the first map of Fig. 2). This was expected since Biella is the most important city in terms of demand.
- 10 municipalities ($p = 10$) by the end of 2022: some small municipalities close to and other big ones far from Biella are chosen (see the second map in Fig. 2).
- 37 municipalities ($p = 37$) by the end of 2025: the solution tends to select municipalities close to the previously selected ones, creating clusters (see the third map in Fig. 2)
- all municipalities ($p = 78$) by the end of 2030 (this corresponds to the trivial solution with $y_i = 1, \forall i \in N$).

The value of all the KPIs, in the various steps of intervention, is calculated and shown in Table 2. Note that the last column, corresponding to the case in which all the locations are chosen, contains the best possible value for each KPI. Several observations can be done:

- D_{max} decreases with the increase in the number of municipalities in which at least one charging station has been located and, as it can be seen, it reaches reasonable values from $p = 10$ onward.
- D^h_{max} increases with the increase in the number of municipalities in which at least one charging station has been located. This KPI is complementary to the D_{max} since it is the importance of the node most distant from a server. Hence, its monotonicity is not a standard feature of the model and it strongly depends on the instance considered.
- D_{min} decreases as the number of municipalities in which at least one charging station has been located increases, and it stabilizes at the best value already with $p = 10$.
- D^h_{min} decreases as the number of municipalities in which at least one charging station has been located increases. This KPI is complementary to the D_{min} since it is the importance of the node nearest to a server. Hence, its monotonicity is not a standard feature of the model and it strongly depends on the instance considered. Usually, the smallest distance is achieved by the internal distance of the node in which an electric station is located. In fact, if just 1 location is considered, the node nearest to a facility is the most important (Biella). Then, if more stations are

Fig. 2 Optimal location for $p = 1$, 10 and 37 (2019). Chosen locations in red

Table 2 KPIs value in the four intervention p-centdian

KPI	$p = 1$ (2019)	$p = 10$ (2022)	$p = 37$ (2025)	$p = 78$ (2030)
D_{max}	53	24	20	11
D^h_{max}	0.001	0.001	0.001	0.002
D_{min}	5.7	2	2	2
D^h_{min}	0.285	0.005	0.005	0.001
D_{avg}	20.3	8.9	5.8	4.4
D^h_{avg}	0.19	0.08	0.06	0.06
$Disp$	5.73	2158.2	34,663.9	167,201.3
Acc	0.024769	0.115986	0.329689	0.456748
C	55%	96%	98%	100%
C^h	0.781	0.993	1	1
RC^h	0.062	4.44	14.32	29.41
C_{min}	0	0	0	1
C^h_{min}	0.002	0.002	0.001	0.001
C_{max}	1	8	23	43
C^h_{max}	0.285	0.285	0.285	0.285
C_{avg}	0.089744	2.653846	8.833333	19.28205
C^h_{avg}	0.01	0.06	0.21	0.38

located, the nearest node has a lower importance, this is reasonable since node with a smaller internal distance usually are less important (they have a smaller population, hence a smaller demand).

- D_{avg} decreases as the number of municipalities in which at least one charging station has been located increases. It is interesting to note that the percentage improvement in the indicator decreases as the number of selected municipalities increases.
- D^h_{avg} decreases as the number of municipalities in which at least one charging station has been located increases. It reaches its asymptotic value already when p=10 (faster than D_{avg}). This is reasonable since the model considers the weighted distances.
- *Disp* increases as the number of municipalities in which at least one charging station has been located increases. Its growth is very marked due to the factorial growth of the number of pairs of selected municipalities. The starting value is set to zero since with a single municipality the summation in the definition cannot be calculated.
- *Acc* increases as the number of municipalities in which at least one charging station has been located increases. Also in this case the improvements are less marked as the number of selected municipalities increases.
- C increases as the number of municipalities in which at least one charging station has been located increases. It can be seen that with only 10 selected municipalities, the coverage reaches very high levels (96% of the municipalities are covered).

- C^h increases as the number of municipalities in which at least one charging station has been located increases. Its convergence is faster than the convergence of C because the nodes that are not covered until $p = 78$ have less importance.
- RC^h increases as the number of municipalities in which at least one charging station has been located increases. It is important to note that the largest part of this KPI is provided by the multiple coverage of the most important nodes. As the reader can notice, the increment of the value is very fast.
- C_{min} increases with the number of municipalities where at least one charging station has been located. Since this is the most pessimistic case, this indicator remains at zero when 1, 10, and 37 selected municipalities are considered. The data then verifies the non-total coverage shown by the KPI previously discussed.
- C^h_{min} decreases as the number of municipalities in which at least one charging station has been located increases. As the reader can notice, the least covered nodes have always a very low importance.
- C_{max} increases as the number of municipalities in which at least one charging station has been located increases. It can be seen that the increase in value grows with the number of selected municipalities. However, it can be noted that already with 10 municipalities the most covered municipality has the choice between 7 charging stations within a 25 km radius.
- C^h_{max} is constant with respect to the number of municipalities in which at least one charging station has been located increases. This is due to the fact that at each iteration, the node that is covered the greatest number of times is the most important node, i.e, the city of Biella.
- C_{avg} increases with the increase in the number of municipalities in which at least one charging station has been located and, as it can be seen, has a much lower value than the C_{max}. This implies a heterogeneous situation in terms of coverage of the various locations. In fact, we have a large number of municipalities covered by a few charging stations and a small number of municipalities covered by many charging stations. Since the towns that are not covered are those with a lower demand (i.e., with less electric vehicles) this feature is in line with the technical specifications of the problem.
- C^h_{avg} increases with the increase of the number of municipalities in which at least one charging station has been located. The increase rate of this value is similar to the one of C_{avg}, except for the first step $p = 1 \rightarrow p = 10$.

A common trend of almost all the KPIs is that the second intervention is the one providing the highest proportional change with respect to the previous one (e.g., C almost doubles its value for $p = 10$ while it gains only few units for $p = 37$ and $p = 78$). Interesting enough, D_{min} reaches its optimal value even for $p = 10$. This represents a very important insight for the company for two main reasons. First, it means that the users will perceive the biggest improvement in terms of service in relatively small amount of time (the first 3–5 years) and in response to a small effort in terms of installed stations. Second, it means that the last interventions, which are the ones affected by the most uncertainty (e.g., in terms of economical sustainability), are not very critical for the process overall quality.

7 Conclusions

In this paper, we have conducted an extensive comparative analysis of models and Key Performance Indicators concerning the optimal location of charging stations for electric vehicles. Motivated by a real case-study concerning the district of Biella in Italy, we highlighted the fact that a perfect location model does not exist but, instead, different models might be jointly considered to face a certain set of requirements and objectives. Therefore, a battery of topological and coverage Key Performance Indicators have been identified and calculated for the solutions given by the different models. The analyzed KPIs include measures about the covering capabilities, the robustness, the dispersion, and the accessibility of the resulting solutions.

Several future research lines can be defined. First, a similar analysis can be performed by explicitly including stochasticity into the decision process. Given the application at hand, the demand rate h_i of node i is the parameter that makes more sense to represent as a stochastic variable. This is due both to the difficulty of estimating the service demand Q_i for any node according to a static vision (e.g., number of EV users living around a demand node) and for the unpredictable dynamics of traffics flows and their issues. Second, as already mentioned in the Introduction, a comparative study can be performed for several other location models.

References

1. Cai, H., Jia, X., Chiu, A.S., Hu, X., Xu, M.: Siting public electric vehicle charging stations in Beijing using big-data informed travel patterns of the taxi fleet. Transp. Res. Part D Transp. Environ. **33**, 39–46 (2014). https://doi.org/10.1016/j.trd.2014.09.003
2. Chen, T., Kockelman, K., Khan, M.: The electric vehicle charging station location problem: a parking-based assignment method for seattle. In: Transportation Research Board 92nd Annual Meeting, pp. 13–1254 (2013)
3. Cherkesly, M., Landete, M., Laporte, G.: Median and covering location problems with interconnected facilities. Comput. Oper. Res. **107**, 1–18 (2019). https://doi.org/10.1016/j.cor.2019.03.002
4. Dashora, Y., Barnes, J., Pillai, R.S., Combs, T.E., Hilliard, M., Sudhan Chinthavali, M.: The PHEV charging infrastructure planning (PCIP) problem. Int. J. Emer. Electr. Power Syst. **11** (2010). https://doi.org/10.2202/1553-779X.2482
5. Fadda, E., Fotio Tiotsop, L., Manerba, D., Tadei, R.: The stochastic multi-path Traveling Salesman Problem with dependent random travel costs. Transp. Sci. (2020). https://doi.org/10.1287/trsc.2020.0996
6. Fadda, E., Gobbato, L., Perboli, G., Rosano, M., Tadei, R.: Waste collection in urban areas: a case study. Interfaces **48**(4), 307–322 (2018)
7. Fadda, E., Manerba, D., Tadei, R., Camurati, P., Cabodi, G.: KPIs for optimal location of charging stations for electric vehicles: the Biella case-study. In: Ganzha, M., Maciaszek, L., Paprzycki M. (eds.) Proceedings of the 2019 Federated Conference on Computer Science and Information Systems. Annals of Computer Science and Information Systems, vol. 18, pp. 123–126. IEEE (2019). https://doi.org/10.15439/2019F171
8. Fadda, E., Perboli, G., Squillero, G.: Adaptive batteries exploiting on-line steady-state evolution strategy. In: Squillero, G., Sim, K. (eds.) Applications of Evolutionary Computation, pp. 329–341. Springer International Publishing, Cham (2017)

9. Fadda, E., Perboli, G., Tadei, R.: Customized multi-period stochastic assignment problem for social engagement and opportunistic IoT. Comput. Oper. Res. **93**, 41–50 (2018)
10. Fadda, E., Perboli, G., Tadei, R.: A progressive hedging method for the optimization of social engagement and opportunistic IoT problems. Eur. J. Oper. Res. **277**(2), 643–652 (2019). https://doi.org/10.1016/j.ejor.2019.02.052. http://www.sciencedirect.com/science/article/pii/S0377221719302164
11. Fadda, E., Perboli, G., Vallesio, V., Mana, D.: Sustainable mobility and user preferences by crowdsourcing data: the open agora project. In: 14th IEEE International Conference on Automation Science and Engineering, CASE 2018, Munich, Germany, 20–24 Aug 2018, pp. 1243–1248 (2018). https://doi.org/10.1109/COASE.2018.8560512
12. Frade, I., Ribeiro, A., Goncalves, G., Pais Antunes, A.: Optimal location of charging stations for electric vehicles in a neighborhood in Lisbon, Portugal. Transp. Res. Rec. J. Transp. Res. Board **2252**, 91–98 (2011). https://doi.org/10.3141/2252-12
13. Frade, I., Ribeiro, A., Goncalves, G., Pais Antunes, A.: An optimization model for locating electric vehicle charging stations in central urban areas. Transp. Res. Rec. J. Transp. Res. Board **3582**, 1–19 (2011)
14. Hansen, W.: How accessibility shapes land use. J. Am. Inst. Plann. **25**, 73–76 (1959). https://doi.org/10.1080/01944365908978307
15. He, J., Yang, H., Tang, T.Q., Huang, H.J.: An optimal charging station location model with the consideration of electric vehicle's driving range. Transp. Res. Part C Emer. Technol. **86**, 641–654 (2018). https://doi.org/10.1016/j.trc.2017.11.026. http://www.sciencedirect.com/science/article/pii/S0968090X17303558
16. Labbé, M., Leal, M., Puerto, J.: New models for the location of controversial facilities: a bilevel programming approach. Comput. Oper. Res. **107**, 95–106 (2019). https://doi.org/10.1016/j.cor.2019.03.003
17. Lee, C., Han, J.: Benders-and-price approach for electric vehicle charging station location problem under probabilistic travel range. Transp. Res. Part B Methodol. **106**, 130–152 (2017). https://doi.org/10.1016/j.trb.2017.10.011. http://www.sciencedirect.com/science/article/pii/S0191261517305052
18. Massiani, J., Gohs, A.: The choice of bass model coefficients to forecast diffusion for innovative products: an empirical investigation for new automotive technologies. Res. Transp. Econ. **50**, 17–28 (2015) (Electric Vehicles: Modelling Demand and Market Penetration). https://doi.org/10.1016/j.retrec.2015.06.003. http://www.sciencedirect.com/science/article/pii/S0739885915000220
19. Quiliot, A., Sarbinowski, A.: Facility location models for vehicle sharing systems. In: Proceedings of the 2016 Federated Conference on Computer Science and Information Systems, vol. 8, pp. 605–608 (2016). https://doi.org/10.15439/2016F10
20. Repower: La mobilità sostenibile e i veicoli elettrici. https://www.repower.com/media/115267/repower-whitepapermobilita-2017_001.pdf (2017)
21. Sadeghi-Barzani, P., Rajabi-Ghahnavieh, A., Kazemi-Karegar, H.: Optimal fast charging station placing and sizing. Appl. Energy **125**, 289–299 (2014). https://doi.org/10.1016/j.apenergy.2014.03.077. http://www.sciencedirect.com/science/article/pii/S0306261914003171
22. Tadei, R., Ricciardi, N., Perboli, G.: The stochastic p-median problem with unknown cost probability distribution. Oper. Res. Lett. **37**, 135–141 (2009). https://doi.org/10.1016/j.orl.2009.01.005
23. Wu, F., Sioshansi, R.: A stochastic flow-capturing model to optimize the location of fast-charging stations with uncertain electric vehicle flows. Transp. Res. Part D Transp. Environ. **53**, 354–376 (2017). https://doi.org/10.1016/j.trd.2017.04.035. http://www.sciencedirect.com/science/article/pii/S136192091630102X
24. Xi, X., Sioshansi, R., Marano, V.: Simulation-optimization model for location of a public electric vehicle charging infrastructure. Transp. Res. Part D Transp. Environ. **22**, 60–69 (2013). https://doi.org/10.1016/j.trd.2013.02.014
25. Zhu, Z.H., Gao, Z.Y., Zheng, J.F., Du, H.M.: Charging station location problem of plug-in electric vehicles. J. Transp. Geogr. **52**, 11–22 (2016). https://doi.org/10.1016/j.jtrangeo.2016.02.002

InterCriteria Analysis of the Evaporation Parameter Influence on Ant Colony Optimization Algorithm: A Workforce Planning Problem

Olympia Roeva, Stefka Fidanova, and Maria Ganzha

Abstract Optimization of the production process is an important task for every factory or organization. A better organization can be done by optimization of the workforce planing. The main goal is decreasing the assignment cost of the workers with the help of which, the work will be done. The problem is NP-hard, therefore it can be solved with algorithms coming from artificial intelligence. The problem is to select employers and to assign them to the jobs to be performed. The constraints of this problem are very strong and it is difficult to find feasible solutions. We apply Ant Colony Optimization Algorithm (ACO) to solve the problem. We investigate the algorithm performance by changing the evaporation parameter. The aim is to find the best parameter setting. To evaluate the influence of the evaporation parameter on ACO InterCriteria Analysis (ICrA) is applied. ICrA is performed on the ACO results for 10 problems considering average and maximum number of iterations needed to solve the problem. Five different values of evaporation parameter are used. The results show that ACO algorithm has best performance for two values of evaporation parameter – 0.1 and 0.3.

Keywords Intercriteria analysis · Workforce planning · Ant colony optimization · Metaheuristics · Evaporation parameter

---

O. Roeva (✉)
Institute of Biophysics and Biomedical Engineering,
Bulgarian Academy of Sciences, Sofia, Bulgaria
e-mail: olympia@biomed.bas.bg

S. Fidanova
Institute of Information and Communication Technology,
Bulgarian Academy of Sciences, Sofia, Bulgaria
e-mail: stefka@parallel.bas.bg

M. Ganzha
System Research Institute, Polish Academy of Sciences, Warsaw and Management Academy,
Warsaw, Poland
e-mail: maria.ganzha@ibspan.waw.pl

S. Fidanova (ed.), *Recent Advances in Computational Optimization*,
Studies in Computational Intelligence 920,
https://doi.org/10.1007/978-3-030-58884-7_5

1 Introduction

The workforce planning is a very important decision making problem for branches of the industry. It plays an important role in human resource management. It includes multiple level of complexity, therefore it is a hard optimization problem (NP-hard). The problem can be divided in to two parts: selection and assignment. The first part is selection of employers from the set of available workers. The second part is assignment of the selected workers to jobs, which the worker will perform. The goal is to carry out the work requirements minimizing assignment cost.

As we have noted above the problem is very hard optimization problem and cannot be solved with exact methods or traditional numerical methods for instances with realistic size. These types of methods can be applied only on simplified variants of the problem. A deterministic version of workforce planing problem is studied in [1, 2]. In [1] the workforce planning is reformulated as mixed integer programming. It is shown that the mixed integer program is much easier to solve the problem than the non-linear program. In [2] the model includes workers differences and the possibility of workers training and upgrading. A variant with random demands of the problem is considered in [3, 4]. Two stage program of scheduling and allocating with random demands is proposed in [3]. Other variant of the problem is to include uncertainty [5–9]. A lot of authors skip some of the constraints to simplify the problem. Mixed linear programming is apply in [10] and in [4] is utilized decomposition method, but for the more complex non-linear workforce planning problems, the convex methods are not applicable.

Last decade, nature-inspired metaheuristic methods receive more and more attention, because they can find close to optimal solutions even for large-scale difficult problems [11–15]. In the literature can be found various metaheuristic algorithms solving workforce planning problem. They include genetic algorithm [16, 17], memetic algorithm [18], scatter search [16] etc.

Ant Colony Optimization (ACO) algorithm is proved to be very effective solving various complex optimization problems [19, 20]. In our previous work [21, 22] we propose ACO algorithm for workforce planning. We have considered the variant of the workforce planning problem proposed in [16]. Current paper is the continuation of [21] and further develops the ideas behind [21]. We investigate the influence of evaporation parameter on algorithm performance.The aim is to find the best parameter setting.

In this paper we present one more possibility to evaluate the influence of the evaporation parameter on algorithm performance, using temporal approaches as Index Matrices and Intuitionistic Fuzzy Sets. We apply the InterCriteria Analysis (ICrA) approach [23]. ICrA aiming to go beyond the nature of the criteria involved in a process of evaluation of multiple objects against multiple criteria, and, thus to discover some dependencies between the ICrA criteria themselves [23].

Initially, ICrA has been applied for the purposes of temporal, threshold and trends analyses of an economic case-study of European Union member states' competitiveness [24, 25]. Further, ICrA has been used to discover the dependencies of different problems as [26–30] and analysis of the performance of some metaheuristics as GAs and ACO [19, 31, 32]. Published results show the applicability of the ICrA and the correctness of the approach.

The rest of the paper is organized as follows. In Sect. 2 the mathematical description of the problem is presented. In Sect. 3 ACO algorithm for workforce planing problem is described. The theoretical background of the ICrA is given in Sect. 4. Section 5 shows computational results, comparisons and discussion. A conclusion and directions for future work are done in Sect. 6.

2 Workforce Planning Problem

In this paper we consider the workforce planning problem proposed in [16] and [33]. The set of jobs $J = \{1, \ldots, m\}$ need to be completed during a fixed period of time. The job j requires d_j hours to be finished. $I = \{1, \ldots, n\}$ is the set of workers, candidates to be assigned. Every worker must perform every of assigned to him job minimum $h_{\min}$ hours can work in efficient way. The worker i is available s_i hours. One worker can be assigned to maximum $j_{\max}$ jobs. The set A_i shows the jobs, that worker i is qualified. Maximum t workers can be assigned during the planed period, or at most t workers may be selected from the set I of workers. The selected workers need to be capable to complete all the jobs they are assigned. The goal is to find feasible solution, that optimizes the objective function.

The cost of assigning the worker i to the job j is c_{ij}. The mathematical model of the workforce planing problem is described as follows:

$$x_{ij} = \begin{cases} 1 \text{ if the worker } i \text{ is assigned to job j} \\ 0 \text{ otherwise} \end{cases}$$

$$y_i = \begin{cases} 1 \text{ if worker } i \text{ is selected} \\ 0 \text{ otherwise} \end{cases}$$

$$z_{ij} = \text{number of hours that worker } i$$

$$\text{is assigned to perform job } j$$

$$Q_j = \text{set of workers qualified to perform job } j$$

$$\text{Minimize} \sum_{i \in I} \sum_{j \in A_i} c_{ij} \cdot x_{ij} \qquad (1)$$

Subject to

$$\sum_{j \in A_i} z_{ij} \leq s_i.y_i \quad i \in I \tag{2}$$

$$\sum_{i \in Q_j} z_{ij} \geq d_j \quad j \in J \tag{3}$$

$$\sum_{j \in A_i} x_{ij} \leq j_{\max} \cdot y_j \quad i \in I \tag{4}$$

$$h_{\min} \cdot x_{ij} \leq z_{ij} \leq s_i \cdot x_{ij} \quad i \in I, j \in A_i \tag{5}$$

$$\sum_{i \in I} y_i \leq t \tag{6}$$

$$\begin{aligned} &x_{ij} \in \{0, 1\} \; i \in I, j \in A_i \\ &y_i \in \{0, 1\} \; i \in I \\ &z_{ij} \geq 0 \qquad i \in I, j \in A_i \end{aligned}$$

The objective function is the minimization of the total assignment cost. The number of hours for each selected worker is limited (inequality 2). The work must be done in full (inequality 3). The number of the jobs, that every worker can perform is limited (inequality 4). There is minimal number of hours that every job must be performed by every assigned worker can work efficiently (inequality 5). The number of assigned workers is limited (inequality 6).

This mathematical model can be used with other objectives too. If $\tilde{c}_{ij}$ is the cost the worker i to performs the job j for one hour, than the objective function can minimize the cost of the hall jobs to be finished.

$$f(x) = \min \sum_{i \in I} \sum_{j \in A_i} \tilde{c}_{ij} \cdot x_{ij} \tag{7}$$

The preferences of the workers to the jobs can be included. In this case one of the variants of the objective function will be to maximize the satisfaction of the workers preferences.

3 Ant Colony Optimization Algorithm

The ACO is a nature inspired methodology. It is a metaheuristics, following the real ants behavior when looking for a food. Real ants use chemical substance, called pheromone, to mark their path ant can return back. An ant moves in random way and when it detects a previously laid pheromone it decides whether to follow it and reinforce it with a new added pheromone. Thus the more ants follow a given trail, the more attractive that trail becomes. There is evaporation in a nature and the pheromone evaporates during the time. Thus the pheromone level of not used and less used paths decreases and they become less desirable. In this way the nature prevents the ants to follow some wrong and useless path. The ants can find a shorter path between the source of the food and the nest by their collective intelligence.

3.1 Main ACO Algorithm

It is not practical to solve HP-hard problems with exact methods or traditional numerical methods when the problem is large. An option is to be applied some metaheuristics. The goal is to find a good solution for a reasonable computational resources like time and memory [34].

For a first time, ant behavior is used for solving optimization problems by Marco Dorigo [35]. Later some modifications are proposed, mainly in pheromone updating rules [34]. The basic in ACO methodology is the simulation of ants behavior. The problem is represented by graph. The solutions are represented by paths in a graph and the aim is to find shorter path corresponding to given constraints. The requirements of ACO algorithm are as follows:

- Appropriate representation of the problem by a graph;
- Appropriate pheromone placement on the nodes or on the arcs of the graph;
- Suitable problem-dependent heuristic function, which manage the ants to improve solutions;
- Pheromone updating rules;
- Transition probability rule, which specifies how to include new nodes in the partial solution;
- Appropriate algorithm parameters.

The transition probability $P_{i,j}$, is a product of the heuristic information $\eta_{i,j}$ and the pheromone trail level $\tau_{i,j}$ related to the move from node i to the node j, where $i, j = 1, \ldots, n$.

$$P_{i,j} = \frac{\tau_{i,j}^a \cdot \eta_{i,j}^b}{\sum\limits_{k \in Unused} \tau_{i,k}^a \cdot \eta_{i,k}^b}, \qquad (8)$$

where $Unused$ is the set of unused nodes of the graph.

The initial pheromone level is the same for all elements of the graph and is set to a positive constant value $\tau_0, 0 < \tau_0 < 1$. After that at the end of the current iteration the ants update the pheromone level [34]. A node become more desirable if it accumulates more pheromone.

The main update rule for the pheromone is:

$$\tau_{i,j} \leftarrow \rho \cdot \tau_{i,j} + \Delta\tau_{i,j}, \qquad (9)$$

where ρ decreases the value of the pheromone, which mimics evaporation in a nature. $\Delta\tau_{i,j}$ is a new added pheromone, which is proportional to the quality of the solution. For measurement of the quality of the solution is used the value of the objective function of the ants solution.

The first node of the solution is randomly chosen. With the random start the search process is diversifying and the number of ants may be small according the number of the nodes of the graph and according other population based metaheuristic methods. The heuristic information represents the prior knowledge of the problem, which is used to better manage the algorithm performance. The pheromone is a global history of the ants to find optimal solution. It is a tool for concentration of the search around best so far solutions.

3.2 *Workforce Planing ACO*

An important role for the successes of the ACO algorithm is the representation of the problem by graph. The graph we propose is 3 dimensional and the node (i, j, z) corresponds to worker i to be assigned to the job j for time z. When an ant begins their tour we generate three random numbers: the first random number is from the interval $[0, \dots, n]$ and corresponds to the worker we assign; the second random number is from the interval $[0, \dots, m]$ and shows the job which this worker will perform. The third random number is from the interval $[h_{\min}, \min\{d_j, s_i\}]$ and shows number of hours worker i is assigned to performs job j. Next node is included in the solution, applying transition probability rule. We repeat this procedure till the solution is constructed.

The following heuristic information is applied:

$$\eta_{ijl} = \begin{cases} l/c_{ij} & l = z_{ij} \\ 0 & otherwise \end{cases} \tag{10}$$

By this heuristic information the most cheapest unassigned worker, is assigned as longer as possible. The node with highest probability from all possible nodes is chosen to be included in the partial solution. When there are more than one possibilities with the same probability, the next node is chosen in a random way between them.

When a new node is included we take in to account all constraints: how many workers are assigned till now; how many time slots every worker is assigned till now; how many time slots are assigned per job till now. If a move do not meets all constraints, the probability of this move is set to 0. The solution is constructed if there are not more possibilities for including new nodes (the transition probability is 0 for all possible moves). If the constructed solution is feasible the value of the objective function is the sum of the assignment cost of the assigned workers. When the constructed solution is not feasible, the value of the objective function is set to be equal to -1.

New pheromone is deposited only on the elements of feasible solutions. The deposited pheromone is proportional to the reciprocal value of the objective function.

$$\Delta\tau_{i,j} = \frac{\rho - 1}{f(x)} \tag{11}$$

Thus the nodes belonging to better solutions accumulate more pheromone than others and will be more attractive in the next iteration. The iteration best solution is compared with the global best solution and if on the current iteration the some of the ants achieves better solution it becomes the new global best. As end condition we use the number of iterations.

In this research we are concentrated on influence of the evaporation parameter on algorithm performance. We tested several values for this parameter and compare the number of needed iterations to find the best solution.

4 InterCriteria Analysis

InterCriteria analysis, based on the apparatuses of index matrices [36–40] and intuitionistic fuzzy sets (IFSs) [41–43], is given in details in [23]. Here, for completeness, the proposed idea is briefly presented.

An intuitionistic fuzzy pair (IFP) [44] is an ordered pair of real non-negative numbers $\langle a, b \rangle$, where $a, b \in [0, 1]$ and $a + b \leq 1$, that is used as an evaluation of some object or process. According to [44], the components (a and b) of IFP might be interpreted as degrees of "membership" and "non-membership" to a given set, degrees of "agreement" and "disagreement", degrees of "validity" and "non-validity", degrees of "correctness" and "non-correctness", etc.

The apparatus of index matrices is presented initially in [37] and discussed in more details in [38, 39]. For the purposes of ICrA application, the initial index set consists of the criteria (for rows) and objects (for columns) with the index matrix elements assumed to be real numbers. Further, an index matrix with index sets consisting of the criteria (for rows and for columns) with IFP elements determining the degrees of correspondence between the respective criteria is constructed, as it is going to be briefly presented below.

Let the initial index matrix is presented in the form of Eq. (12), where, for every p, q, $(1 \leq p \leq m, 1 \leq q \leq n)$, C_p is a criterion, taking part in the evaluation; O_q – an object to be evaluated; $C_p(O_q)$ – a real number (the value assigned by the p-th criteria to the q-th object).

$$A = \begin{array}{c|ccccccc} & O_1 & \ldots & O_k & \ldots & O_l & \ldots & O_n \\ C_1 & C_1(O_1) & \ldots & C_1(O_k) & \ldots & C_1(O_l) & \ldots & C_1(O_n) \\ \vdots & \vdots & \ddots & \vdots & \ddots & \vdots & \ddots & \vdots \\ C_i & C_i(O_1) & \ldots & C_i(O_k) & \ldots & C_i(O_l) & \ldots & C_i(O_n) \\ \vdots & \vdots & \ddots & \vdots & \ddots & \vdots & \ddots & \vdots \\ C_j & C_j(O_1) & \ldots & C_j(O_k) & \ldots & C_j(O_l) & \ldots & C_j(O_n) \\ \vdots & \vdots & \ddots & \vdots & \ddots & \vdots & \ddots & \vdots \\ C_m & C_m(O_1) & \ldots & C_m(O_k) & \ldots & C_m(O_l) & \ldots & C_m(O_n) \end{array} . \quad (12)$$

Let O denotes the set of all objects being evaluated, and $C(O)$ is the set of values assigned by a given criteria C (i.e., $C = C_p$ for some fixed p) to the objects, i.e.,

$$O \stackrel{\text{def}}{=} \{O_1, O_2, O_3, \ldots, O_n\},$$
$$C(O) \stackrel{\text{def}}{=} \{C(O_1), C(O_2), C(O_3), \ldots, C(O_n)\}.$$

Let $x_i = C(O_i)$. Then the following set can be defined:

$$C^*(O) \stackrel{\text{def}}{=} \{\langle x_i, x_j\rangle | i \neq j \,\&\, \langle x_i, x_j\rangle \in C(O) \times C(O)\}.$$

Further, if $x = C(O_i)$ and $y = C(O_j)$, $x \prec y$ iff $i < j$ will be written.

In order to find the agreement of different criteria, the vectors of all internal comparisons for each criterion are constructed, which elements fulfil one of the three relations R, $\overline{R}$ and $\tilde{R}$. The nature of the relations is chosen such that for a fixed criterion C and any ordered pair $\langle x, y\rangle \in C^*(O)$:

$$\langle x, y\rangle \in R \Leftrightarrow \langle y, x\rangle \in \overline{R}, \quad (13)$$
$$\langle x, y\rangle \in \tilde{R} \Leftrightarrow \langle x, y\rangle \notin (R \cup \overline{R}), \quad (14)$$
$$R \cup \overline{R} \cup \tilde{R} = C^*(O). \quad (15)$$

For example, if “R” is the relation “$<$”, then $\overline{R}$ is the relation “$>$”, and vice versa.

Hence, for the effective calculation of the vector of internal comparisons (denoted further by $V(C)$) only the considering of a subset of $C(O) \times C(O)$ is needed, namely:

$$C^{\prec}(O) \stackrel{\text{def}}{=} \{\langle x, y\rangle |\, x \prec y \,\&\, \langle x, y\rangle \in C(O) \times C(O),$$

due to Eqs. (13)–(15). For brevity, $c^{i,j} = \langle C(O_i), C(O_j)\rangle$.

Then for a fixed criterion C the vector of lexicographically ordered pair elements is constructed:

$$V(C) = \{c^{1,2}, c^{1,3}, \ldots, c^{1,n}, c^{2,3}, c^{2,4}, \ldots, c^{2,n}, c^{3,4}, \ldots, c^{3,n}, \ldots, c^{n-1,n}\}. \quad (16)$$

In order to be more suitable for calculations, $V(C)$ is replaced by $\hat{V}(C)$, where its k-th component ($1 \leq k \leq \frac{n(n-1)}{2}$) is given by:

$$\hat{V}_k(C) = \begin{cases} 1, \text{ iff } V_k(C) \in R, \\ -1, \text{ iff } V_k(C) \in \overline{R}, \\ 0, \text{ otherwise.} \end{cases}$$

When comparing two criteria the degree of "agreement" is determined as the number of matching components of the respective vectors (divided by the length of the vector for normalization purposes). This can be done in several ways, e.g. by counting the matches or by taking the complement of the Hamming distance. The degree of "disagreement" is the number of components of opposing signs in the two vectors (again normalized by the length).

If the respective degrees of "agreement" and "disagreement" are denoted by $\mu_{C,C'}$ and $\nu_{C,C'}$, it is obvious (from the way of computation) that $\mu_{C,C'} = \mu_{C',C}$ and $\nu_{C,C'} = \nu_{C',C}$. Also it is true that $\langle \mu_{C,C'}, \nu_{C,C'} \rangle$ is an IFP.

In the most of the obtained pairs $\langle \mu_{C,C'}, \nu_{C,C'} \rangle$, the sum $\mu_{C,C'} + \nu_{C,C'}$ is equal to 1. However, there may be some pairs, for which this sum is less than 1. The difference

$$\pi_{C,C'} = 1 - \mu_{C,C'} - \nu_{C,C'} \tag{17}$$

is considered as a degree of uncertainty.

The following index matrix is constructed as a result of applying the ICrA to A (Eq. (12)):

$$\begin{array}{c|ccc} & C_2 & \dots & C_m \\ \hline C_1 & \langle \mu_{C_1,C_2}, \nu_{C_1,C_2} \rangle & \dots & \langle \mu_{C_1,C_m}, \nu_{C_1,C_m} \rangle \\ \vdots & \vdots & \ddots & \vdots \\ C_{m-1} & & \dots & \langle \mu_{C_{m-1},C_m}, \nu_{C_{m-1},C_m} \rangle \end{array},$$

that determines the degrees of correspondence between criteria C_1, ..., C_m.

In this paper we use μ-biased algorithm **Algorithm 1** for calculation of intercriteria relations [45]. An example pseudocode of the **Algorithm 1** is presented below.

5 Computational Results and Discussion

5.1 Numerical Experiments

In this section we tested our algorithm on 10 structured problems in order to investigate evaporation parameter influence. The software, which realizes the algorithm

Algorithm 1 Calculating $\mu_{C,C'}$ and $\nu_{C,C'}$ between two criteria

Require: Vectors $\hat{V}(C)$ and $\hat{V}(C')$

```
1: function DEGREES OF AGREEMENT AND DISAGREEMENT(V̂(C), V̂(C'))
2:     V ← V̂(C) − V̂(C')
3:     μ ← 0
4:     ν ← 0
5:     for i ← 1 to n(n−1)/2 do
6:         if V_i = 0 then
7:             μ ← μ + 1
8:         else if abs(V_i) = 2 then          ▷ abs(V_i): the absolute value of V_i
9:             ν ← ν + 1
10:        end if
11:    end for
12:    μ ← 2/(n(n−1)) μ
13:    ν ← 2/(n(n−1)) ν
14:    return μ, ν
15: end function
```

```
Ant Colony Optimization
Initialize number of ants;
Initialize the ACO parameters;
while not end-condition do
    for k = 0 to number of ants
        ant k choses start node;
        while solution is not constructed do
            ant k selects higher probability node;
        end while
    end for
    Update-pheromone-trails;
end while
```

Fig. 1 Pseudocode for ACO

is written in C and is run on Pentium desktop computer at 2.8 GHz with 4 GB of memory. The structure of the ACO algorithm is shown by the pseudocode in Fig. 1.

An artificially generated problem instances considered in [16] is used for the tests. The test instances characteristics are shown in Table 1.

The parameter settings of our ACO algorithm are shown in Table 2. The ACO parameters are fixed experimentally after several runs of the algorithm.

In a previous work [21] we show that presented here ACO algorithm outperforms the genetic algorithm and scatter search algorithm proposed in [16]. We perform 30 independent runs with every one of the five values of the evaporation parameter (Table 2), because the algorithm is stochastic and to guarantee the robustness of the average results. We apply ANOVA test for statistical analysis to guarantee the significance of the difference between the average results. We compare the average number of iterations needed to find the best result for every test problem. The needed num-

Table 1 Test instances characteristics

Parameters	Value
n	20
m	20
t	10
s_i	[50, 70]
$j_{\max}$	[27, 31]
$h_{\min}$	[38, 44]

Table 2 ACO parameter settings

Parameters	Value
Number of iterations	100
ρ	$\{0.1, 0.3, 0.5, 0.7, 0.9\}$
τ_0	0.5
Number of ants	20
a	1
b	1

Table 3 Evaporation parameter ranking

	$\rho_1 = 0.1$	$\rho_2 = 0.3$	$\rho_3 = 0.5$	$\rho_4 = 0.7$	$\rho_5 = 0.9$
First place	3 times	4 times	1 times	2 times	0 times
Second place	2 times	3 times	3 times	2 times	1 times
Third place	2 times	1 times	3 times	2 times	1 times
Forth place	2 times	1 times	2 times	3 times	2 times
Fifth plase	1 times	1 times	1 times	1 times	6 times
Ranking	26	22	29	29	43

ber of iterations for every test problem can be very different, because the specificity of the tests. Therefore for comparison we use ranking as more representative. The algorithm with some fixed value for evaporation is on the first place, if it achieves the best solution with less average number of iterations over 30 runs, according other values and we assign to it 1. We assign 2 to the value on the second place, 3 to the value on the third place, 4 to the value of the forth place and 5 to the value with the larger number of iterations. For some cases it can be assigned the same numbers if the number of iterations to find the best solution is the same. We sum the ranking of the cases over all 10 test problems in order to find final ranking of ACO with the different values of the evaporation parameter. The ranking results are shown in Table 3.

The less number of iterations is needed when the evaporation parameter is equal to 0.3. In this case the algorithm is on the first place four times, on the second place is 3 times and on the third, fourth and fifth—one time, respectively. The worst results are achieved when the value of the evaporation parameter is 0.9. The results achieved when the evaporation parameter is 0.1 are a little bit worse than when the value of the evaporation parameter is equal to 0.3. When the value of the evaporation parameter increases, the results obtained become worse.

5.2 Application of the InterCriteria Analysis

In this section we use ICrA to obtain some additional knowledge about the influence of the evaporation parameter on the ACO performance. Based on the obtained results we construct the following index matrices:

- IM with the average number of iterations (Eq. (18))

Problem	$\rho_1 = 0.1$	$\rho_2 = 0.3$	$\rho_3 = 0.5$	$\rho_4 = 0.7$	$\rho_5 = 0.9$
$S20_{01}$	10.46	14.56	17	20.13	17.53
$S20_{02}$	34.16	32.96	34.16	34.73	49.46
$S20_{03}$	51.83	58.4	44.13	31.16	60.16
$S20_{04}$	163.53	321.16	143	79.6	229.93
$S20_{05}$	13.03	7.96	13	8.7	10.2
$S20_{06}$	15.6	16.46	17	15.9	18.63
$S20_{07}$	74.23	47.13	75	66.83	60.93
$S20_{08}$	156.93	101	88.13	123	209.93
$S20_{09}$	59.13	66.46	83.1	85.1	94.33
$S20_{10}$	16.83	16.76	19	22	35

(18)

- IM with the maximum number of iterations (Eq. (19))

Problem	$\rho_1 = 0.1$	$\rho_2 = 0.3$	$\rho_3 = 0.5$	$\rho_4 = 0.7$	$\rho_5 = 0.9$
$S20_{01}$	24	30	39	61	61
$S20_{02}$	65	38	65	65	66
$S20_{03}$	94	94	94	78	93
$S20_{04}$	207	355	308	306	355
$S20_{05}$	80	24	83	30	63
$S20_{06}$	32	32	33	47	48
$S20_{07}$	143	143	144	143	143
$S20_{08}$	224	224	224	224	225
$S20_{09}$	100	170	170	117	169
$S20_{10}$	47	47	46	54	104

(19)

We applied ICrA for the IMs Eqs. (18) and (19), using the software ICrAData [46] (see Fig. 2).

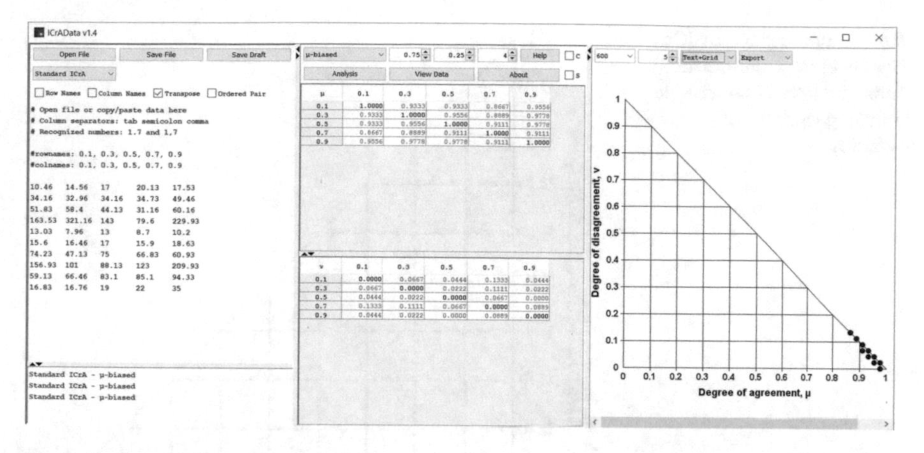

Fig. 2 Screen shot form ICrAData

As a result we obtained the two index matrices with the relations between considered five values of the evaporation parameter for both cases—average and maximum iteration number.

5.3 *ICrA Results of Data for Average Number of Iterations*

Based on the IM from Eq. (18) the resulting index matrices for $\mu_{C,C'}$, $\nu_{C,C'}$ and $\pi_{C,C'}$ values are shown in Eqs. (20), (21) and (22), respectively.

$$\begin{array}{c|ccccc} \mu_{C,C'} & \rho_1 & \rho_2 & \rho_3 & \rho_4 & \rho_5 \\ \rho_1 & 1 & 0.93 & 0.93 & 0.87 & 0.96 \\ \rho_2 & 0.93 & 1 & 0.96 & 0.89 & 0.98 \\ \rho_3 & 0.93 & 0.96 & 1 & 0.91 & 0.98 \\ \rho_4 & 0.87 & 0.89 & 0.91 & 1 & 0.91 \\ \rho_5 & 0.96 & 0.98 & 0.98 & 0.91 & 1 \end{array} \tag{20}$$

$$\begin{array}{c|ccccc} \nu_{C,C'} & \rho_1 & \rho_2 & \rho_3 & \rho_4 & \rho_5 \\ \rho_1 & 0 & 0.07 & 0.04 & 0.13 & 0.04 \\ \rho_2 & 0.07 & 0 & 0.02 & 0.11 & 0.02 \\ \rho_3 & 0.04 & 0.02 & 0 & 0.07 & 0.00 \\ \rho_4 & 0.13 & 0.11 & 0.07 & 0 & 0.09 \\ \rho_5 & 0.04 & 0.02 & 0.00 & 0.09 & 0 \end{array} \tag{21}$$

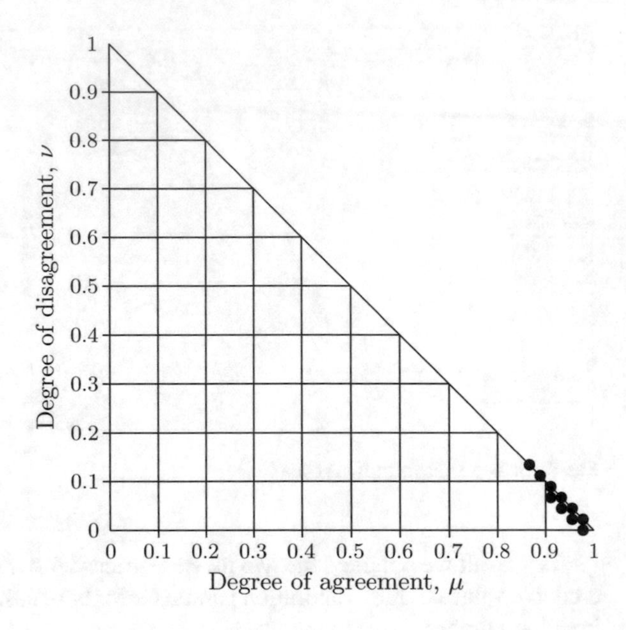

Fig. 3 Presentation of ICrA results in the intuitionistic fuzzy interpretation triangle – average number of iterations

$\pi_{C,C'}$	ρ_1	ρ_2	ρ_3	ρ_4	ρ_5
ρ_1	0	0	0.02	0	0
ρ_2	0	0	0.02	0	0
ρ_3	0.02	0.02	0	0.02	0.02
ρ_4	0	0	0.02	0	0
ρ_5	0	0	0.02	0	0

(22)

The obtained ICrA results are visualized within the specific triangular geometrical interpretation of IFSs on Fig. 3 [47].

5.4 ICrA Results of Data for Maximum Number of Iterations

Based on the IM from Eq. (19) the resulting index matrices for $\mu_{C,C'}$, $\nu_{C,C'}$ and $\pi_{C,C'}$ values are shown in Eqs. (23), (24) and (25), respectively.

$\mu_{C,C'}$	ρ_1	ρ_2	ρ_3	ρ_4	ρ_5
ρ_1	1	0.84	0.93	0.84	0.84
ρ_2	0.84	1	0.87	0.91	0.91
ρ_3	0.93	0.87	1	0.87	0.91
ρ_4	0.84	0.91	0.87	1	0.87
ρ_5	0.84	0.91	0.91	0.87	1

(23)

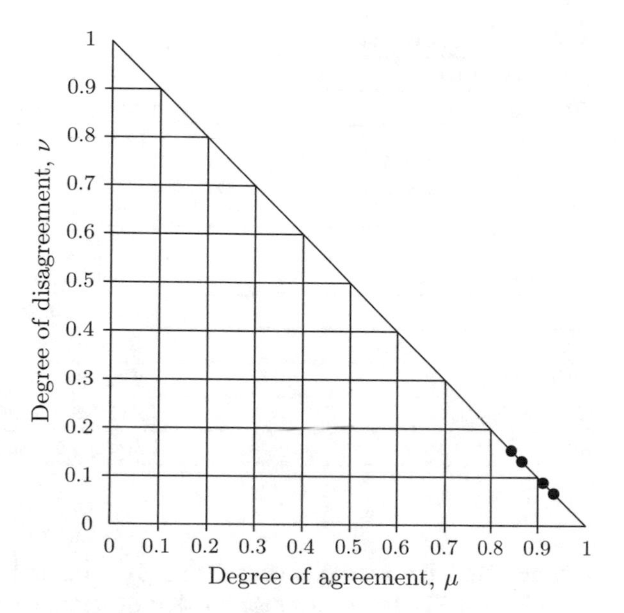

Fig. 4 Presentation of ICrA results in the intuitionistic fuzzy interpretation triangle—maximum number of iterations

$\nu_{C,C'}$	ρ_1	ρ_2	ρ_3	ρ_4	ρ_5
ρ_1	0	0.16	0.07	0.16	0.16
ρ_2	0.16	0	0.13	0.09	0.09
ρ_3	0.07	0.13	0	0.13	0.09
ρ_4	0.16	0.09	0.13	0	0.13
ρ_5	0.16	0.09	0.09	0.13	0

(24)

$\pi_{C,C'}$	ρ_1	ρ_2	ρ_3	ρ_4	ρ_5
ρ_1	0	0	0	0	0
ρ_2	0	0	0	0	0
ρ_3	0	0	0	0	0
ρ_4	0	0	0	0	0
ρ_5	0	0	0	0	0

(25)

The obtained ICrA results are visualized within the specific triangular geometrical interpretation of IFSs on Fig. 4.

The ICrA results are analysed based on the proposed in [48] consonance and dissonance scale. The scheme for defining the consonance and dissonance between each pair of criteria is presented in Table 4.

The obtained relations between the algorithms' performance based on different values of evaporation parameter are mainly in positive consonance and three results are in strong positive consonance ($\rho_1 - \rho_5$, $\rho_2 - \rho_5$ and $\rho_3 - \rho_5$). This means that there is no essential difference between the algorithms' performance. Such a result leads to the conclusion that the data used in this particular form are uninformative. The number of iterations observed for the 10 test problems examined varied over

Table 4 Consonance and dissonance scale [48]

Interval of $\mu_{C,C'}$	Meaning
[0–0.05]	Strong negative consonance (SNC)
(0.05–0.15]	Negative consonance (NC)
(0.15–0.25]	Weak negative consonance (WNC)
(0.25–0.33]	Weak dissonance (WD)
(0.33–0.43]	Dissonance (D)
(0.43–0.57]	Strong dissonance (SD)
(0.57–0.67]	Dissonance (D)
(0.67–0.75]	Weak dissonance (WD)
(0.75–0.85]	Weak positive consonance (WPC)
(0.85–0.95]	Positive consonance (PC)
(0.95–1]	Strong positive consonance (SPC)

a wide range. For example, considering $\rho_2 = 0.3$ the number of iterations varies from 7.96 to 321 for the average number of iterations and from 24 to 355 for the maximum number of iterations. Due to the specific nature of the problems, it is not appropriate to use the data in this way. To obtain more representative data and as a more appropriate approach, data from the ranking of the results obtained by the number of iterations should be used.

5.5 ICrA Results of Data from Ranking of the Results

We construct the following IM Eq. (26):

$$\begin{array}{c|ccccc} & 1stplace & 2ndplace & 3rdpalce & 4thplace & 5thplace \\ \rho_1 & 3 & 2 & 2 & 2 & 1 \\ \rho_2 & 4 & 3 & 1 & 1 & 1 \\ \rho_3 & 1 & 3 & 3 & 2 & 1 \\ \rho_4 & 2 & 2 & 2 & 3 & 1 \\ \rho_5 & 0 & 1 & 1 & 2 & 6 \end{array} \quad (26)$$

Based on the IM Eq. (26) ICrA is applied. The obtained results are presented as index matrices for $\mu_{C,C'}$, $\nu_{C,C'}$ and $\pi_{C,C'}$ values, shown in Eqs. (27), (28) and (29), respectively.

$$
\begin{array}{c|ccccc}
\mu_{C,C'} & \rho_1 & \rho_2 & \rho_3 & \rho_4 & \rho_5 \\
\rho_1 & 1 & 1.00 & 0.50 & 0.80 & 0.00 \\
\rho_2 & \mathbf{1.00} & 1 & 0.40 & 0.50 & 0.00 \\
\rho_3 & 0.50 & 0.40 & 1 & 0.67 & 0.37 \\
\rho_4 & \mathbf{0.80} & 0.50 & 0.67 & 1 & 0.43 \\
\rho_5 & \mathbf{0.00} & \mathbf{0.00} & 0.37 & 0.43 & 1
\end{array}
\tag{27}
$$

$$
\begin{array}{c|ccccc}
\nu_{C,C'} & \rho_1 & \rho_2 & \rho_3 & \rho_4 & \rho_5 \\
\rho_1 & 0 & 0.00 & 0.50 & 0.20 & 1.00 \\
\rho_2 & 0.00 & 0 & 0.60 & 0.50 & 1.00 \\
\rho_3 & 0.50 & 0.60 & 0 & 0.33 & 0.62 \\
\rho_4 & 0.20 & 0.50 & 0.33 & 0 & 0.57 \\
\rho_5 & 1.00 & 1.00 & 0.62 & 0.57 & 0
\end{array}
\tag{28}
$$

$$
\begin{array}{c|ccccc}
\pi_{C,C'} & \rho_1 & \rho_2 & \rho_3 & \rho_4 & \rho_5 \\
\rho_1 & 0 & 0 & 0 & 0 & 0 \\
\rho_2 & 0 & 0 & 0 & 0 & 0 \\
\rho_3 & 0 & 0 & 0 & 0 & 0.01 \\
\rho_4 & 0 & 0 & 0 & 0 & 0 \\
\rho_5 & 0 & 0 & 0.01 & 0 & 0
\end{array}
\tag{29}
$$

The obtained ICrA results are visualized on Fig. 5 within the specific triangular geometrical interpretation of IFSs.

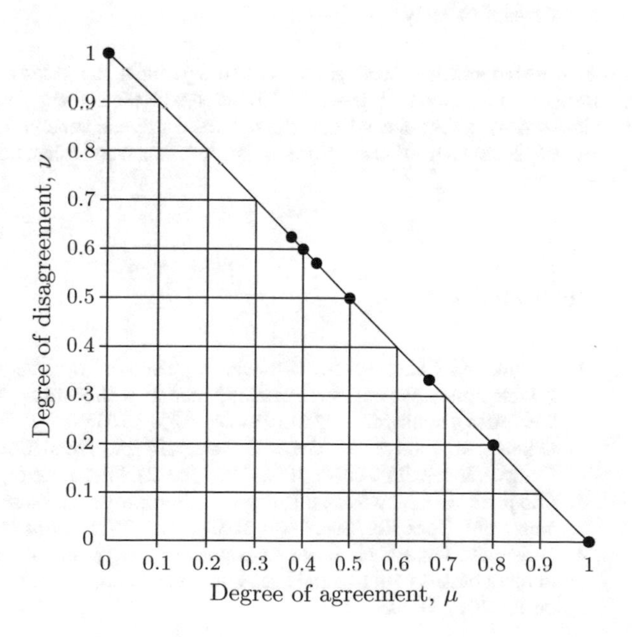

Fig. 5 Presentation of ICrA results in the intuitionistic fuzzy interpretation triangle— parameter ranking

Analyzing the estimated relations between considered five values of the evaporation parameter based on the results' ranking, we found that ACO algorithm with $\rho = 0.1$ and $\rho = 0.3$ perform identically. The obtained relationship between these to ACO algorithm is strong positive consonance. ACO algorithm with $\rho = 0.9$ shows strong negative consonance compared to the ACO with $\rho = 0.1$ and $\rho = 0.3$. This means in all cases when the ACO algorithm with $\rho = 0.1$ (or ACO algorithm with $\rho = 0.3$) performs well, the ACO algorithm with $\rho = 0.9$ performs do not performs well.

In summary, ICrA shows that the ACO algorithm performs is the same for both $\rho = 0.1$ and $\rho = 0.3$.

6 Conclusion

In this paper we apply ACO algorithm to solve workforce planning problem. We are concentrated on the influence of the evaporation parameter on the algorithm performance, how many iterations are needed to find the best solution. We test the algorithm on 10 structured benchmark problems. The achieved results show that when the value of the evaporation parameter is small, the algorithm needs less number of iterations compared with high value of the evaporation parameter. The numerical calculations show that the results achieved when the evaporation parameter is 0.1 are a little bit worse than when the value of the evaporation parameter is 0.3. Further performed ICrA shows that the both ACO algorithms (with $\rho = 0.1$ and $\rho = 0.3$) performs similarly.

Acknowledgements Work presented here is partially supported by the National Scientific Fund of Bulgaria under Grant KP-06-N22/1 "Theoretical Research and Applications of InterCriteria Analysis" and by the European Union through the European structural and Investment funds Grant No BG05M2OP001-1.001-0003, financed by the Science and Education for Smart Growth Operational Program.

References

1. Hewitt, M., Chacosky, A., Grasman, S., Thomas, B.: Integer programming techniques for solving non-linear workforce planning models with learning. Euro. J. Oper. Res. **242**(3), 942–950 (2015). https://doi.org/10.1016/j.ejor.2014.10.060
2. Othman, M., Bhuiyan, N., Gouw, G.: Integrating workers' differences into workforce planning. Comput. Indus. Eng. **63**(4), 1096–1106 (2012). https://doi.org/10.1016/j.cie.2012.06.015
3. Campbell, G.: A two-stage stochastic program for scheduling and allocating cross-trained workers. J. Oper. Res. Soc. **62**(6), 1038–1047 (2011). https://doi.org/10.1057/jors.2010.16
4. Parisio, A., Jones, C.N.: A two-stage stochastic programming approach to employee scheduling in retail outlets with uncertain demand. Omega **53**, 97–103 (2015). https://doi.org/10.1016/j.omega.2015.01.003

5. Hu, K., Zhang, X., Gen, M., Jo, J.: A new model for single machine scheduling with uncertain processing time. J. Intell. Manufact. **28**(3), 717–725 (2015). https://doi.org/10.1007/s10845-015-1033-9
6. Li, R., Liu, G.: An uncertain goal programming model for machine scheduling problem. J. Intel. Manuf. **28**(3), 689–694 (2014). https://doi.org/10.1007/s10845-014-0982-8
7. Ning, Y., Liu, J., Yan, L.: Uncertain aggregate production planning. Soft Comput. **17**(4), 617–624 (2013). https://doi.org/10.1007/s00500-012-0931-4
8. Yang, G., Tang, W., Zhao, R.: An uncertain workforce planning problem with job satisfaction. Int. J. Machine Learn. Cybern. 2016 (Springer). https://doi.org/10.1007/s13042-016-0539-6http://rd.springer.com/article/10.1007/s13042-016-0539-6
9. Zhou, C., Tang, W., Zhao, R.: An uncertain search model for recruitment problem with enterprise performance. J Intell. Manufact. **28**(3), 295–704 (2014). https://doi.org/10.1007/s10845-014-0997-1
10. Easton, F.: Service completion estimates for cross-trained workforce schedules under uncertain attendance and demand. Prod. Oper. Manage. **23**(4), 660–675 (2014). https://doi.org/10.1111/poms.12174
11. Albayrak, G., Zdemir, I.: A state of art review on metaheuristic methods in time-cost trade-off problems. Int. J. Structu. Civil Eng. Res. **6**(1), 30–34 (2017). https://doi.org/10.18178/ijscer.6.1.30-34
12. Mucherino, A., Fidanova, S., Ganzha, M.: Introducing the environment in ant colony optimization, recent advances in computational optimization, studies in computational. Intelligence **655**, 147–158 (2016). https://doi.org/10.1007/978-3-319-40132-4_9
13. Roeva, O., Atanassova, V.: Cuckoo search algorithm for model parameter identification. Int. J. Bioautomation **20**(4), 483–492 (2016)
14. Tilahun, S.L., Ngnotchouye, J.M.T.: Firefly algorithm for discrete optimization problems: a survey. J. Civil Eng. **21**(2), 535–545 (2017). https://doi.org/10.1007/s12205-017-1501-1
15. Toimil, D., Gmes, A.: Review of metaheuristics applied to heat exchanger network design. Int. Trans. Oper. Res. **24**(1–2), 7–26 (2017). https://doi.org/10.1111/itor.12296
16. Alba, E., Luque, G., Luna, F.: Parallel metaheuristics for workforce planning. J. Math. Modell. Algorithm. **6**(3), 509–528 (2007). https://doi.org/10.1007/s10852-007-9058-5
17. Li, G., Jiang, H., He, T.: A genetic algorithm-based decomposition approach to solve an integrated equipment-workforce-service planning problem. Omega **50**, 1–17 (2015). https://doi.org/10.1016/j.omega.2014.07.003
18. Soukour, A., Devendeville, L., Lucet, C., Moukrim, A.: A Memetic algorithm for staff scheduling problem in airport security service. Expert Syst. Appl. **40**(18), 7504–7512 (2013). https://doi.org/10.1016/j.eswa.2013.06.073
19. Fidanova, S., Roeva, O., Paprzycki, M., Gepner, P.: InterCriteria Analysis of ACO Start Startegies. In: Proceedings of the 2016 Federated Conference on Computer Science and Information Systems, 2016, pp. 547-550. https://doi.org/10.1007/978-3-319-99648-6_4
20. Grzybowska, K., Kovcs, G.: Sustainable supply chain—supporting tools. In: Proceedings of the 2014 Federated Conference on Computer Science and Information Systems, vol. 2, 2014, pp. 1321–1329. https://doi.org/10.15439/2014F75
21. Fidanova, S., Luquq, G., Roeva, O., Paprzycki, M., Gepner, P.: Ant colony optimization algorithm for workforce planning. In: FedCSIS'2017, IEEE Xplorer, IEEE Catalog Number CFP1585N-ART, 2017, pp. 415–419. https://doi.org/10.15439/2017F63
22. Roeva, O., Fidanova, S., Luque, G., Paprzycki, M., Gepner, P.: Hybrid ant colony optimization algorithm for workforce planning. In: FedCSIS'2018. IEEE Xplorer, pp. 233–236 (2018). https://doi.org/10.15439/2018F47
23. Atanassov, K., Mavrov, D., Atanassova, V.: Intercriteria decision making: a new approach for multicriteria decision making, based on index matrices and intuitionistic fuzzy sets. Issues Intuitionistic Fuzzy Sets Generalized Nets **11**, 1–8 (2014)
24. Atanassova, V., Mavrov, D., Doukovska, L., Atanassov, K.: Discussion on the threshold values in the intercriteria decision making approach. Notes Intuitionistic Fuzzy Sets **20**(2), 94–99 (2014)

25. Atanassova, V., Doukovska, L., Atanassov, K., Mavrov, D.: Intercriteria decision making approach to EU member states competitiveness analysis. In: Proceedings of the International Symposium on Business Modeling and Software Design— BMSD'14, pp. 289–294 (2014)
26. Antonov, A.: Dependencies between model indicators of general and special speed in 13–14 year old hockey players. Trakia J. 2020. (in press)
27. Antonov, A.: Analysis and detection of the degrees and direction of correlations between key indicators of physical fitness of 10–12-year-old hockey players. Int. J. Bioautomation **23**(3), 303–314 (2019). https://doi.org/10.7546/ijba.2019.23.3.000709
28. Todinova, S., Mavrov, D., Krumova, S., Marinov, P., Atanassova, V., Atanassov, K., Taneva, S.G.: Blood plasma thermograms dataset analysis by means of intercriteria and correlation analyses for the case of colorectal cancer. Int. J. Bioautomation **20**(1), 115–124 (2016)
29. Vassilev, P., Todorova, L., Andonov, V.: An auxiliary technique for InterCriteria Analysis via a three dimensional index matrix. Notes Intuitionistic Fuzzy Sets **21**(2), 71–76 (2015)
30. Zaharieva, B., Doukovska, L., Ribagin, S., Radeva, I.: InterCriteria decision making approach for behterev's disease analysis. Int. J. Bioautomation **24**(1), 5–14 (2020). https://doi.org/10.7546/ijba.2020.24.1.000507
31. Angelova, M., Roeva, O., Pencheva, T.: InterCriteria analysis of crossover and mutation rates relations in simple genetic algorithm. In: Proceedings of the 2015 Federated Conference on Computer Science and Information Systems, Vol. 5, pp. 419–424 (2015)
32. Roeva, O., Fidanova, S., Vassilev, P., Gepner, P.: InterCriteria analysis of a model parameters identification using genetic algorithm. Proce. Federated Conf. Comput. Sci. Inf. Syst. **5**, 501–506 (2015)
33. Glover, F., Kochenberger, G., Laguna, M., Wubbena, T.: Selection and assignment of a skilled workforce to meet job requirements in a fixed planning period. In: MAEB'04, 2004, pp. 636–641
34. Dorigo, M., Stutzle, T.: Ant Colony Optimization. MIT Press (2004)
35. Bonabeau, E., Dorigo, M., Theraulaz, G.: Swarm Intelligence: From Natural to Artificial Systems. Oxford University Press, New York (1999)
36. Atanassov, K.: Index Matrices: Towards an Augmented Matrix Calculus. Springer, Switzerland (2014)
37. Atanassov, K.: Generalized index matrices. Comptes rendus de l'Academie bulgare des Sciences **40**(11), 15–18 (1987)
38. Atanassov, K.: On index matrices, part 1: standard cases. Adv. Stud. Contemp. Math. **20**(2), 291–302 (2010)
39. Atanassov, K.: On index matrices, part 2: intuitionistic fuzzy case. Proce. Jangjeon Math. Soc. **13**(2), 121–126 (2010)
40. Atanassov, K.: On index matrices. Part 5: 3-dimensional index matrices. Adv. Stud. Contemp. Math. **24**(4), 423–432 (2014)
41. Atanassov, K.: Intuitionistic fuzzy sets. VII ITKR session, Sofia, 20–23 June 1983. (Reprinted) Int. J. Bioautomation, **20**(S1), S1–S6 (2016)
42. Atanassov, K.: On Intuitionistic Fuzzy Sets Theory. Springer, Berlin (2012)
43. Atanassov, K.: Review and new results on intuitionistic fuzzy sets, mathematical foundations of artificial intelligence seminar, Sofia, 1988, Preprint IM-MFAIS-1-88. (Reprinted) Int. J. Bioautomation **20**(S1), S7–S16 (2016)
44. Atanassov, K., Szmidt, E., Kacprzyk, J.: On intuitionistic fuzzy pairs. Notes Intuitionistic Fuzzy Sets **19**(3), 1–13 (2013)
45. Roeva, O., Vassilev, P., Angelova, M., Su, J., Pencheva, T.: Comparison of different algorithms for InterCriteria relations calculation. In: 2016 IEEE 8th International Conference on Intelligent Systems, pp. 567–572 (2016)
46. Ikonomov, N., Vassilev, P., Roeva, O.: ICrAData software for intercriteria analysis. Int. J. Bioautomation **22**(1), 1–10 (2018)

47. Atanassova, V.: Interpretation in the intuitionistic fuzzy triangle of the results, obtained by the intercriteria analysis. In: Proceedings of the 9th Conference of the European Society for Fuzzy Logic and Technology (EUSFLAT), pp. 1369–1374 (2015)
48. Atanassov, K., Atanassova, V., Gluhchev, G.: Inter criteria analysis: ideas and problems. Notes Intuitionistic Fuzzy Sets **21**(1), 81–88 (2015)

Caterpillar Alignment Distance for Rooted Labeled Caterpillars: Distance Based on Alignments Required to Be Caterpillars

Yoshiyuki Ukita, Takuya Yoshino, and Kouichi Hirata

Abstract A *rooted labeled caterpillar* (*caterpillars*, for short) is a rooted labeled tree transformed to a rooted path after removing all the leaves in it. In this paper, we discuss the *alignment distance* between two caterpillars, which is formulated as the minimum cost of possible alignments as supertrees of them. First, we point out that, whereas the alignment between two trees is always a tree, the alignment between two caterpillars is not always a caterpillar. Then, we formulate a *caterpillar alignment distance* such that the alignment is required to be a caterpillar. Also we formulate a *caterpillar less-constrained mapping* and show that the caterpillar alignment distance coincides with the minimum cost of possible caterpillar less-constrained mappings. Furthermore, we design the algorithm to compute the caterpillar alignment distance between two caterpillars in $O(h^2\lambda^3)$ time under the general cost function and in $O(h^2\lambda)$ time under the unit cost function, where h is the maximum height and λ is the maximum number of leaves in caterpillars. Finally, we give experimental results of computing the caterpillar alignment distance.

Keywords Rooted labeled caterpillar · Alignment distance · Caterpillar alignment distance · Edit distance

---

Y. Ukita (✉)
Graduate School of Computer Science and Systems Engineering,
Kyushu Institute of Technology, Kawazu 680-4, Iizuka 820-8502, Japan
e-mail: o231014y@mail.kyutech.jp

T. Yoshino · K. Hirata
Department of Artificial Intelligence, Kyushu Institute of Technology,
Kawazu 680-4, Iizuka 820-8502, Japan
e-mail: yoshino@ai.kyutech.ac.jp

K. Hirata
e-mail: hirata@ai.kyutech.ac.jp

S. Fidanova (ed.), *Recent Advances in Computational Optimization*,
Studies in Computational Intelligence 920,
https://doi.org/10.1007/978-3-030-58884-7_6

1 Introduction

Comparing tree-structured data such as HTML and XML data for web mining or RNA and glycan data for bioinformatics is one of the important tasks for data mining. The most famous distance measure [3] between *rooted labeled unordered trees* (*trees*, for short) is the *edit distance* [13]. The edit distance is formulated as the minimum cost of *edit operations*, consisting of a *substitution*, a *deletion* and an *insertion*, applied to transform a tree to another tree. It is known that the edit distance is always a metric and coincides with the minimum cost of *Tai mappings* [13].

Unfortunately, the problem of computing the edit distance between trees is MAX SNP-hard [18]. This statement also holds even if trees are binary or the maximum height of trees is at most 3 [1, 5].

Many variations of the edit distance have developed as more structurally sensitive distances (*cf.*, [8, 16]). Almost variations are metrics and the problem of computing them is tractable as cubic-time computable [15–17]. In particular, the *isolated-subtree distance* (or *constrained distance*) [17] is the most general tractable variation of the edit distance [16].

On the other hand, a *caterpillar* (*cf.* [4]) is a tree transformed to a rooted path after removing all the leaves in it. Whereas the caterpillars are very restricted and simple, there are some cases containing many caterpillars in real dataset, see Table 1 in Sect. 6. Recently, Muraka et al. [10] have proposed the algorithm to compute the edit distance between caterpillars in $O(h^2\lambda^3)$ time under the general cost function and in $O(h^2\lambda)$ time under the unit cost function, where h is the maximum height and λ is the maximum number of leaves in caterpillars. They have also introduced the efficient comparable distances to approximate the edit distance between caterpillars [11].

An *alignment distance* is an alternative distance measure between trees, introduced by Jiang et al. [7]. The alignment distance between two trees is formulated as the minimum cost of possible *alignments* (as trees) obtained by first inserting vertices labeled with spaces into two trees so that the resulting trees have the same structure and then overlaying them. In operational, the alignment distance is regarded as an edit distance such that every insertion precedes to deletions. Hence, the alignment distance between trees is not always equal to the edit distance and regarded as a variation of the edit distance. Furthermore, Kuboyama [8] has shown that the alignment distance coincides with the minimum cost of *less-constrained mappings* [9], which is the restriction of the Tai mapping.

As same as the edit distance, the problem of computing the alignment distance between trees is also MAX SNP-hard [7]. On the other hand, it is tractable if the degrees are bounded by some constant [7]. Since a caterpillar is not a bounded-degree tree, it is still open whether or not the problem of computing the alignment distance is tractable. Hence, in this paper, we discuss such a problem.

First of all, we point out that there exists a pair of caterpillars whose minimum cost less-constrained mapping is not an isolated-subtree mapping and whose minimum cost Tai mapping is not a less-constrained mapping. Then, we can apply the algorithm of computing neither the isolated-subtree distance or its variations [15–17, 19] nor the edit distance [10] to compute the alignment distance between caterpillars.

Next, when we apply the standard definition of the alignment between trees [7] to caterpillars, an alignment between caterpillars is not always a caterpillar, whereas an alignment between trees is always a tree. Also there exist two caterpillars such that the minimum cost alignment between them is not a caterpillar. On the other hand, it is required that the alignment between caterpillars is a super-caterpillar of them, as the alignment between trees is a super-tree of them [7].

Then, in this paper, we improve the definitions of the alignment, called a *caterpillar alignment distance*, and the less-constrained mapping between caterpillars, called a *caterpillar less-constrained mapping*, adequate to caterpillars. Then, we show that the caterpillar alignment distance between two caterpillars coincides with the caterpillar less-constrained distance as the minimum cost of all the caterpillar less-constrained mappings, and the caterpillar alignment distance is incomparable with the isolated-subtree distance.

Furthermore, we design the algorithm to compute the alignment distance between caterpillars in $O(h^2\lambda^3)$ time under the general cost function and in $O(h^2\lambda)$ time under the unit cost function. Here, it is necessary to adopt the edit distance for multisets (*cf.*, [11]) to compute the alignment distance between sets of leaves. The time complexity is same as that of computing the edit distance between caterpillars [10, 11].

Finally, we give experimental results of computing the caterpillar alignment distance between caterpillars, by using caterpillars in real dataset in Table 1 in Sect. 6. In particular, we compare the caterpillar alignment distance with the edit distance and the isolated-subtree distance.

2 Preliminaries

In this section, we prepare the notions necessary to discuss the later sections.

A *tree* T is a connected graph (V, E) without cycles, where V is the set of vertices and E is the set of edges. We denote V and E by $V(T)$ and $E(T)$. The *size* of T is $|V|$ and denoted by $|T|$. We sometime denote $v \in V(T)$ by $v \in T$. We denote an empty tree $(\emptyset, \emptyset)$ by $\emptyset$. A *rooted tree* is a tree with one vertex r chosen as its *root*. We denote the root of a rooted tree T by $r(T)$.

Let T be a rooted tree such that $r = r(T)$ and $u, v, w \in T$. We denote the unique path from r to v, that is, the tree (V', E') such that $V' = \{v_1, \ldots, v_k\}$, $v_1 = r$, $v_k = v$ and $(v_i, v_{i+1}) \in E'$ for every i $(1 \leq i \leq k-1)$, by $UP_r(v)$. The *parent* of $v(\neq r)$, which we denote by $par(v)$, is its adjacent vertex on $UP_r(v)$ and the *ancestors* of $v(\neq r)$ are the vertices on $UP_r(v) - \{v\}$. We say that u is a *child* of v if v is the parent of u and u is a *descendant* of v if v is an ancestor of u. We denote the set of children of v by $ch(v)$. We call a vertex with no children a *leaf* and denote the set of all the leaves in T by $lv(T)$.

Let T be a rooted tree (V, E) and v a vertex in T. A *complete subtree of T at v*, denoted by $T[v]$, is a rooted tree $T' = (V', E')$ such that $r(T') = v$, $V' = \{u \in V \mid$

$u \leq v\}$ and $E' = \{(u, w) \in E \mid u, w \in V'\}$. The *degree* of v, denoted by $d(v)$, is the number of children of v, and the *degree* of T, denoted by $d(T)$, is $\max\{d(v) \mid v \in T\}$. The *height* of v, denoted by $h(v)$, is $\max\{|UP_v(w)| \mid w \in lv(T[v])\}$, and the *height* of T, denoted by $h(T)$, is $\max\{h(v) \mid v \in T\}$.

We use the ancestor orders $<$ and $\leq$, that is, $u < v$ if v is an ancestor of u and $u \leq v$ if $u < v$ or $u = v$. We say that w is the *least common ancestor* of u and v, denoted by $u \sqcup v$, if $u \leq w$, $v \leq w$ and there exists no vertex $w' \in T$ such that $w' \leq w$, $u \leq w'$ and $v \leq w'$. When neither $u \leq v$ nor $v \leq u$, we call that *u is incomparable with v*.

For $v \in T$, $pre(v)$ (*resp.*, $post(v)$) denotes the occurrence order of v in the preorder (*resp.*, postorder) traversal of all the vertices in T. Then, we say that u is *to the left of* v in T if $pre(u) \leq pre(v)$ and $post(u) \leq post(v)$. We say that a rooted tree is *ordered* if a left-to-right order among siblings is given; *unordered* otherwise. We say that a rooted tree is *labeled* if each vertex is assigned a symbol from a fixed finite alphabet Σ. For a vertex v, we denote the label of v by $l(v)$, and sometimes identify v with $l(v)$. In this paper, we call a rooted labeled unordered tree a *tree* simply. Furthermore, we call a set of trees a *forest*.

A *rooted path* P is a rooted tree $(\{v_1, \dots, v_n\}, \{(v_i, v_{i+1}) \mid 1 \leq i \leq n-1\})$ such that $r(P) = v_1$. As the restricted form of trees, we introduce a *rooted labeled caterpillar* (*caterpillar*, for short) as follows, which this paper mainly deals with.

Definition 1 (Caterpillar (*cf.*, [4])) We say that a tree is a *caterpillar* if it is transformed to a rooted path after removing all the leaves in it. For a caterpillar C, we call the remained rooted path a *backbone* of C and denote it (and the set of vertices in it) by $bb(C)$.

It is obvious that $r(C) = r(bb(C))$ and $V(C) = bb(C) \cup lv(C)$ for a caterpillar C, that is, every vertex in a caterpillar is either a leaf or an element of the backbone.

Next, we introduce an *edit distance* and a *Tai mapping* between trees.

Definition 2 (Edit operations for trees [13]) The *edit operations* of a tree T are defined as follows, see Fig. 1.

1. *Substitution*: Change the label of the vertex v in T.
2. *Deletion*: Delete a vertex v in T with parent v', making the children of v become the children of v'. The children are inserted in the place of v as a subset of the children of v'. In particular, if v is the root in T, then the result applying the deletion is a forest consisting of the children of the root.
3. *Insertion*: The complement of deletion. Insert a vertex v as a child of v' in T making v the parent of a subset of the children of v'.

Let $\varepsilon \notin \Sigma$ denote a special *blank* symbol and define $\Sigma_\varepsilon = \Sigma \cup \{\varepsilon\}$. Then, we represent each edit operation by $(l_1 \mapsto l_2)$, where $(l_1, l_2) \in (\Sigma_\varepsilon \times \Sigma_\varepsilon - \{(\varepsilon, \varepsilon)\})$. The operation is a substitution if $l_1 \neq \varepsilon$ and $l_2 \neq \varepsilon$, a deletion if $l_2 = \varepsilon$, and an insertion if $l_1 = \varepsilon$. For vertices v and w, we also denote $(l(v) \mapsto l(w))$ by $(v \mapsto w)$.

We define a *cost function* $\gamma : (\Sigma_\varepsilon \times \Sigma_\varepsilon \setminus \{(\varepsilon, \varepsilon)\}) \mapsto \mathbf{R}^+$ on pairs of labels. For $(v, w) \in V(T_1) \times V(T_2)$, we also denote $\gamma(l(v), l(w))$ by $\gamma(v, w)$ simply.

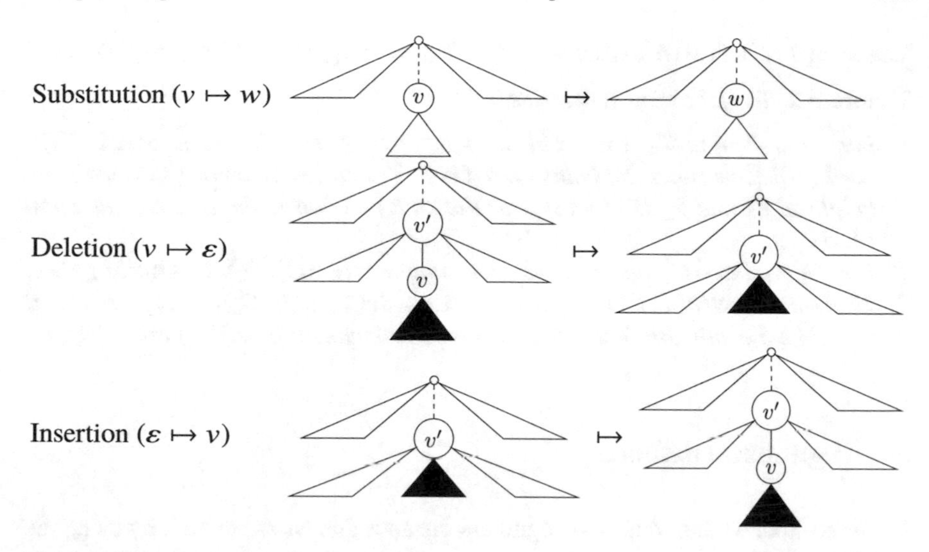

Fig. 1 Edit operations for trees

We often constrain a cost function γ to be a *metric*, that is, $\gamma(l_1, l_2) \geq 0$, $\gamma(l_1, l_2) = 0$ iff $l_1 = l_2$, $\gamma(l_1, l_2) = \gamma(l_2, l_1)$ and $\gamma(l_1, l_3) \leq \gamma(l_1, l_2) + \gamma(l_2, l_3)$. In particular, we call the cost function that $\gamma(l_1, l_2) = 1$ if $l_1 \neq l_2$ a *unit cost function*.

Definition 3 (Edit distance for trees [13]) For a cost function γ, the *cost* of an edit operation $e = l_1 \mapsto l_2$ is given by $\gamma(e) = \gamma(l_1, l_2)$. The *cost* of a sequence $E = e_1, \ldots, e_k$ of edit operations is given by $\gamma(E) = \sum_{i=1}^{k} \gamma(e_i)$. Then, an *edit distance* $\tau_{\mathrm{TAI}}(T_1, T_2)$ between trees T_1 and T_2 is defined as follows:

$$\tau_{\mathrm{TAI}}(T_1, T_2) = \min\{\gamma(E) \mid E \text{ is a sequence of edit operations transforming } T_1 \text{ to } T_2\}.$$

Definition 4 (Tai mapping [13]) Let T_1 and T_2 be trees. We say that a triple (M, T_1, T_2) is a *Tai mapping* (a *mapping*, for short) from T_1 to T_2 if $M \subseteq V(T_1) \times V(T_2)$ and every pair (v_1, w_1) and (v_2, w_2) in M satisfies the following conditions.

1. $v_1 = v_2$ iff $w_1 = w_2$ (one-to-one condition).
2. $v_1 \leq v_2$ iff $w_1 \leq w_2$ (ancestor condition).

We will use M instead of (M, T_1, T_2) when there is no confusion denote it by $M \in \mathcal{M}_{\mathrm{TAI}}(T_1, T_2)$. Furthermore, we denote the set $\{v \in T_1 \mid (v, w) \in M\}$ by $M|_1$ and the set $\{w \in T_2 \mid (v, w) \in M\}$ by $M|_2$.

Let M be a mapping from T_1 to T_2. Then, the *cost* $\gamma(M)$ of M is given as follows.

$$\gamma(M) = \sum_{(v,w)\in M} \gamma(v, w) + \sum_{v\in M|_1} \gamma(v, \varepsilon) + \sum_{w\in M|_2} \gamma(\varepsilon, w).$$

Trees T_1 and T_2 are *isomorphic*, denoted by $T_1 \equiv T_2$, if there exists a mapping $M \in \mathcal{M}_{\mathrm{TAI}}(T_1, T_2)$ such that $M|_1 = M|_2 = \emptyset$ and $\gamma(M) = 0$.

Theorem 1 (Tai [13]) *It holds that* $\tau_{\mathrm{TAI}}(T_1, T_2) = \min\{\gamma(M) \mid M \in \mathcal{M}_{\mathrm{TAI}}(T_1, T_2)\}$.

Theorem 2 *The following statements hold.*

1. *For trees* T_1 *and* T_2*, the problem of computing* $\tau_{\mathrm{TAI}}(T_1, T_2)$ *is* MAX SNP-*hard [18]. This statement holds even if both* T_1 *and* T_2 *are binary, the maximum height of* T_1 *and* T_2 *is at most* 3 *or the cost function is the unit cost function [1, 5].*
2. *For caterpillars* C_1 *and* C_2*, we can compute* $\tau_{\mathrm{TAI}}(C_1, C_2)$ *in* $O(h^2\lambda^3)$ *time, where* $h = \max\{h(C_1), h(C_2)\}$ *and* $\lambda = \max\{|lv(C_1)|, |lv(C_2)|\}$*. Furthermore, if we adopt the unit cost function, then we can compute it in* $O(h^2\lambda)$ *time [10].*

3 Alignment Distance

In this section, we introduce the alignment distance and characterize it by using the variation of Tai mappings.

Definition 5 (Alignment [7]) Let T_1 and T_2 be trees. Then, an *alignment* between T_1 and T_2 is a tree $\mathcal{T}$ obtained by the following steps.

1. Insert new vertices labeled by ε into T_1 and T_2 so that the resulting trees T_1' and T_2' are isomorphic with ignoring labels and $l(\phi(v)) \neq \varepsilon$ whenever $l(v) = \varepsilon$ for an isomorphism ϕ from T_1' to T_2' and every vertex $v \in T_1'$.
2. Set $\mathcal{T}$ to a tree T_1' obtained by relabeling a label $l(v)$ for every vertex $v \in T_1'$ with $(l(v), l(\phi(v)))$. (Note that $(\varepsilon, \varepsilon) \notin \mathcal{T}$.)

Let $\mathcal{A}(T_1, T_2)$ denote the set of all possible alignments between trees T_1 and T_2.

For a cost function γ, the *cost* of an alignment $\mathcal{T}$, denoted by $\gamma(\mathcal{T})$, is the sum of the costs of all labels in $\mathcal{T}$.

Definition 6 (Alignment distance [7]) Let T_1 and T_2 be trees and γ a cost function. Then, an *alignment distance* $\tau_{\mathrm{ALN}}(T_1, T_2)$ between T_1 and T_2 is defined as follows.

$$\tau_{\mathrm{ALN}}(T_1, T_2) = \min\{\gamma(\mathcal{T}) \mid \mathcal{T} \in \mathcal{A}(T_1, T_2)\}.$$

Also we call an alignment between T_1 and T_2 with the minimum cost an *optimal alignment* and denote it by $\mathcal{A}^*(T_1, T_2)$.

The notion of the alignment can be easily extended to forests. The only change is that it is now possible to insert a vertex (as the root) of trees in the forest. We denote the set of all possible alignments between forests F_1 and F_2 by $\mathcal{A}(F_1, F_2)$ and an optimal alignment by $\mathcal{A}^*(F_1, F_2)$.

Example 1 For two caterpillars C_1 and C_2 illustrated in Fig. 2, $\mathcal{A}^*(C_1, C_2)$ is the optimal alignment between C_1 and C_2. Also, for two caterpillars C_3 and C_4 illustrated in Fig. 2, $\mathcal{A}^*(C_3, C_4)$ is the optimal alignment between C_3 and C_4. Under the unit cost function, it holds that $\tau_{\mathrm{ALN}}(C_1, C_2) = 3$ and $\tau_{\mathrm{ALN}}(C_3, C_4) = 3$.

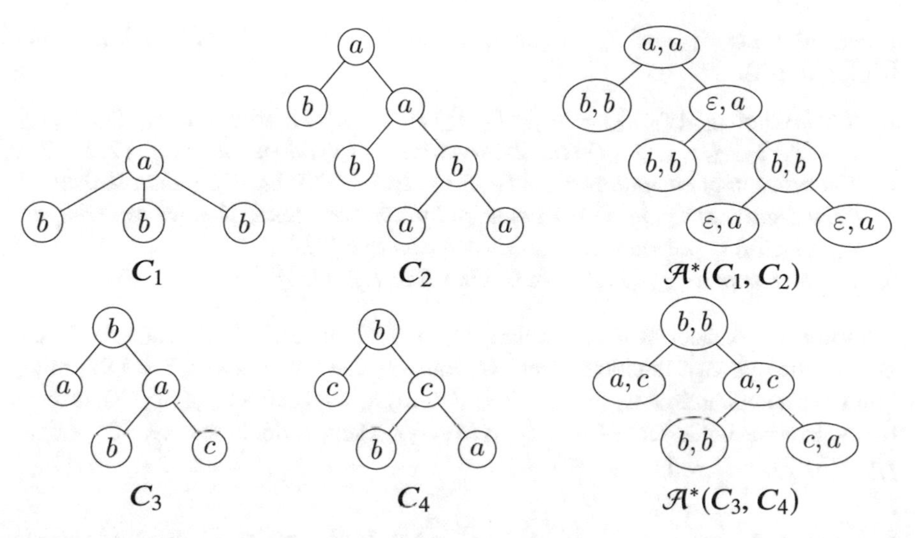

Fig. 2 Caterpillars C_1, C_2, C_3 and C_4 and the optimal alignments $\mathcal{A}^*(C_1, C_2)$ and $\mathcal{A}^*(C_1, C_2)$. in Example 1

Next, we introduce the variations of Tai mappings, including the mapping characterizing the alignment distance.

Definition 7 (Variations of Tai mapping) Let T_1 and T_2 be trees and $M \in \mathcal{M}_{\text{TAI}}(T_1, T_2)$.

1. We say that M is a *less-constrained mapping* [9], denoted by $M \in \mathcal{M}_{\text{LESS}}(T_1, T_2)$, if M satisfies the following condition for every $(v_1, w_1), (v_2, w_2), (v_3, w_3) \in M$:

$$(v_1 \sqcup v_2 < v_1 \sqcup v_3) \Longrightarrow (w_2 \sqcup w_3 = w_1 \sqcup w_3).$$

   Also we define a *less-constrained distance* $\tau_{\text{LESS}}(T_1, T_2)$ as the minimum cost of all the less-constrained mappings, that is:

$$\tau_{\text{LESS}}(T_1, T_2) = \min\{\gamma(M) \mid M \in \mathcal{M}_{\text{LESS}}(T_1, T_2)\}.$$

2. We say that M is an *isolated-subtree mapping* [14] (or a *constrained mapping* [17]), denoted by $M \in \mathcal{M}_{\text{ILST}}(T_1, T_2)$, if M satisfies the following condition for every $(v_1, w_1), (v_2, w_2), (v_3, w_3) \in M$:

$$(v_3 < v_1 \sqcup v_2) \Longleftrightarrow (w_3 < w_1 \sqcup w_2).$$

   Also we define an *isolated-subtree distance* $\tau_{\text{ILST}}(T_1, T_2)$ as the minimum cost of all the isolated-subtree mappings, that is:

$$\tau_{\text{ILST}}(T_1, T_2) = \min\{\gamma(M) \mid M \in \mathcal{M}_{\text{ILST}}(T_1, T_2)\}.$$

Theorem 3 *Let T_1 and T_2 be trees, where $n = \max\{|T_1|, |T_2|\}$ and $d = \min\{d(T_1), d(T_2)\}$.*

1. *It holds that $\tau_{\mathrm{ALN}}(T_1, T_2) = \tau_{\mathrm{LESS}}(T_1, T_2)$ [6, 8]. Also it holds that $\tau_{\mathrm{TAI}}(T_1, T_2) \leq \tau_{\mathrm{ALN}}(T_1, T_2) \leq \tau_{\mathrm{ILST}}(T_1, T_2)$ but the equations always do not hold (cf., [7, 8, 17]).*
2. *The problem of computing $\tau_{\mathrm{ALN}}(T_1, T_2)$ is* MAX SNP*-hard. On the other hand, if the degrees of T_1 and T_2 are bounded by some constants, then we can compute $\tau_{\mathrm{ALN}}(T_1, T_2)$ in polynomial time with respect to n [7].*
3. *We can compute $\tau_{\mathrm{ILST}}(T_1, T_2)$ in $O(n^2 d)$ time (cf., [15]).*

Example 2 Consider two caterpillars C_1 and C_2 in Fig. 2 in Example 1 and assume the unit cost function. Then, M_1 and M_2 illustrated in Fig. 3 are the minimum cost mappings in $\mathcal{M}_{\mathrm{LESS}}(C_1, C_1)(= \mathcal{M}_{\mathrm{TAI}}(C_1, C_2))$ and $\mathcal{M}_{\mathrm{ILST}}(C_1, C_2)$, respectively. Here, it holds that $M_1 \notin \mathcal{M}_{\mathrm{ILST}}(C_1, C_2)$. Then, it holds that $\tau_{\mathrm{TAI}}(C_1, C_2) = \tau_{\mathrm{ALN}}(C_1, C_2) = 3 < 5 = \tau_{\mathrm{ILST}}(C_1, C_2)$.

Example 3 Consider two caterpillars C_3 and C_4 in Fig. 2 in Example 1 and assume the unit cost function. Then, M_3 and M_4 illustrated in Fig. 4 are the minimum cost mappings in $\mathcal{M}_{\mathrm{TAI}}(C_3, C_4)$ and $\mathcal{M}_{\mathrm{LESS}}(C_3, C_4)$. Here, it holds that $M_3 \notin \mathcal{M}_{\mathrm{LESS}}(C_3, C_4)$. Hence, it holds that $\tau_{\mathrm{TAI}}(C_3, C_4) = 2 < 3 = \tau_{\mathrm{ALN}}(C_3, C_4)$.

Example 2 shows that there exists a pair of caterpillars whose minimum cost caterpillar less-constrained mapping is not an isolated-subtree mapping. Then, we cannot use the algorithm to compute the isolated-subtree distance between caterpillars [15–17, 19] to compute their alignment distance. Also Example 3 shows that there exists a pair of caterpillars whose minimum cost Tai mapping is not a caterpillar

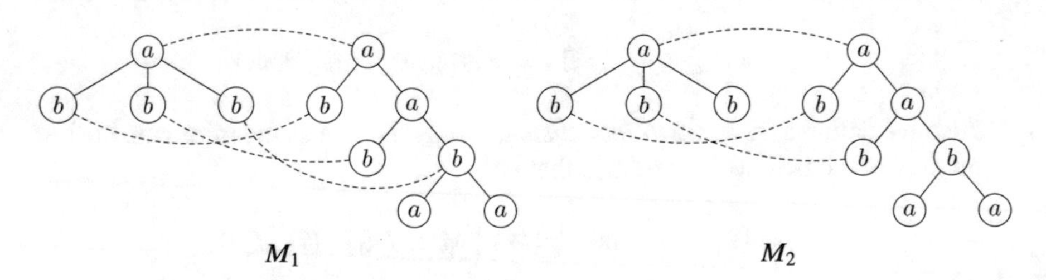

Fig. 3 The minimum cost mappings $M_1 \in \mathcal{M}_{\mathrm{LESS}}(C_1, C_2)$ and $M_2 \in \mathcal{M}_{\mathrm{ILST}}(C_1, C_2)$ in Example 2

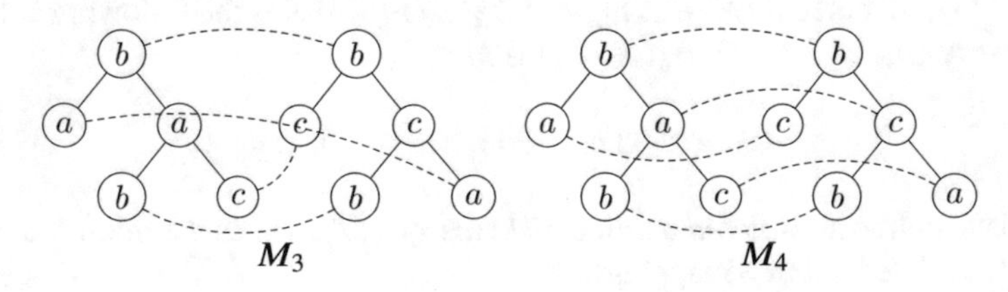

Fig. 4 The minimum cost mappings $M_3 \in \mathcal{M}_{\mathrm{TAI}}(C_3, C_4)$ and $M_4 \in \mathcal{M}_{\mathrm{LESS}}(C_3, C_4)$ in Example 3

less-constrained mapping. Then, we cannot use the algorithm to compute the edit distance between caterpillars [10] to compute their caterpillar alignment distance. Furthermore, it still remains open whether or not Theorem 3 holds for caterpillars.

4 Caterpillar Alignment Distance

In this section, first we point out that there exist two caterpillars such that the minimum cost alignment between them is not a caterpillar.

Example 4 For two caterpillars C_5 and C_6 illustrated in Fig. 5, $\mathcal{A}^*(C_5, C_6)$ is the optimal alignment between C_5 and C_6 and $M_5 \in \mathcal{M}_{\mathrm{LESS}}(C_5, C_6)$ is the corresponding minimum cost less-constrained mapping. Hence, it holds that $\tau_{\mathrm{ALN}}(C_5, C_6) = 2$.
On the other hand, it is obvious that $\mathcal{A}^*(C_5, C_6)$ is not a caterpillar.

In order to avoid the problem in Example 4, we improve the alignment, the alignment distance and the less-constrained mapping for trees to a *caterpillar alignment*, a *caterpillar alignment distance* and a *less-constrained caterpillar mapping*, respectively, for caterpillars.

Definition 8 (Caterpillar alignment)Let C_1 and C_2 be caterpillars. Then, we say that the alignment $\mathcal{A}(C_1, C_2)$ between C_1 and C_2 is a *caterpillar alignment* if it is a caterpillar. We denote the set of all possible caterpillar alignments by $\mathcal{CA}(C_1, C_2)$.

Definition 9 (Caterpillar alignment distance)For a cost function γ, a *caterpillar alignment distance* $\tau_{\mathrm{CALN}}(C_1, C_2)$ between C_1 and C_2 is defined as follows.

$$\tau_{\mathrm{CALN}}(C_1, C_2) = \min\{\gamma(\mathcal{T}) \mid \mathcal{T} \in \mathcal{CA}(C_1, C_2)\}.$$

Also we call a caterpillar alignment between C_1 and C_2 with the minimum cost an *optimal caterpillar alignment* and denote it by $\mathcal{CA}^*(C_1, C_2)$.

Example 5 For caterpillars C_5 and C_6 in Example 4, Fig. 6 (left) illustrates the optimal caterpillar alignment $\mathcal{CA}^*(C_5, C_6)$ between C_5 and C_6. Hence, under the unit cost function, it holds that $\tau_{\mathrm{CALN}}(C_5, C_6) = 3$.

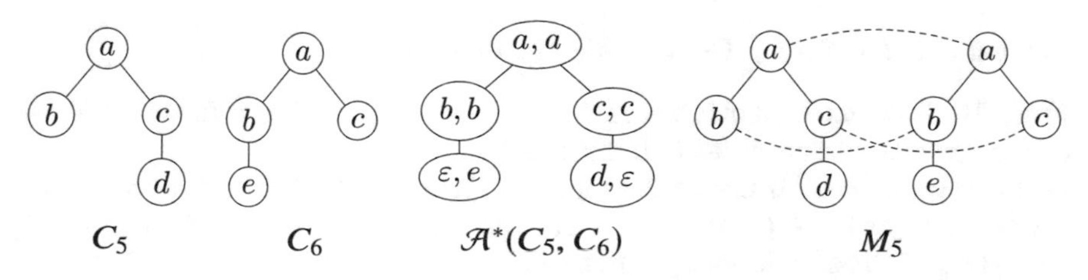

Fig. 5 Caterpillars C_5 and C_6, the optimal alignment $\mathcal{A}^*(C_5, C_6)$ and the minimum cost less-constrained mapping $M_5 \in \mathcal{M}_{\mathrm{LESS}}(C_5, C_6)$ in Example 4

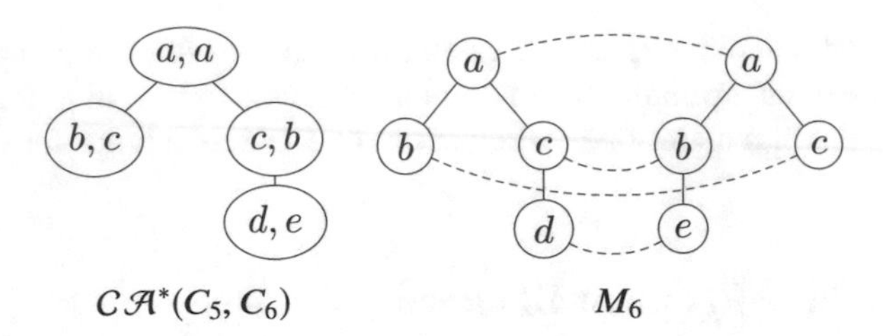

Fig. 6 The optimal caterpillar alignment $\mathcal{CA}^*(C_5, C_6)$ in Example 5 and the minimum cost caterpillar less-constrained mapping $M_6 \in \mathcal{M}_{\text{CLESS}}(C_5, C_6)$ between C_5 and C_6 in Example 6

Definition 10 (Caterpillar less-constrained mapping) Let C_1 and C_2 be caterpillars and $M \in \mathcal{M}_{\text{TAI}}(C_1, C_2)$. Then, we say that M is a *caterpillar less-constrained mapping*, denoted by $M \in \mathcal{M}_{\text{CLESS}}(C_1, C_2)$, if $M \in \mathcal{M}_{\text{LESS}}(C_1, C_2)$ and M contains no pairs (v_1, w_1) and (v_2, w_2) such that $(v_1, w_1) \in bb(C_1) \times lv(C_2)$ and $(v_2, w_2) \in lv(C_1) \times bb(C_2)$. Also we define a *caterpillar less-constrained distance* $\tau_{\text{CLESS}}(C_1, C_2)$ as the minimum cost of all the caterpillar less-constrained mappings, that is:

$$\tau_{\text{CLESS}}(C_1, C_2) = \min\{\gamma(M) \mid M \in \mathcal{M}_{\text{CLESS}}(C_1, C_2)\}.$$

Example 6 For $M_1 \in \mathcal{M}_{\text{LESS}}(C_1, C_2)$ in Fig. 3 and $M_4 \in \mathcal{M}_{\text{LESS}}(C_3, C_4)$ in Fig. 4, it holds that $M_1 \in \mathcal{M}_{\text{CLESS}}(C_1, C_2)$ and $M_4 \in \mathcal{M}_{\text{CLESS}}(C_3, C_4)$, so it holds that $\tau_{\text{CLESS}}(C_1, C_2) = 3$ and $\tau_{\text{CLESS}}(C_3, C_4) = 3$.

On the other hand, for $M_5 \in \mathcal{M}_{\text{LESS}}(C_5, C_6)$ in Fig. 5, it holds that $(b, b) \in lv(C_5) \times bb(C_6)$ and $(c, c) \in bb(C_5) \times lv(C_6)$. Then, it holds that $M_5 \notin \mathcal{M}_{\text{CLESS}}(C_5, C_6)$. Also $M_6 \in \mathcal{M}_{\text{CLESS}}(C_5, C_6)$ in Fig. 6 (right) is the minimum cost caterpillar less-constrained mapping between C_5 and C_6. Hence, under the unit cost function, it holds that $\tau_{\text{CLESS}}(C_5, C_6) = 3$.

According to Theorem 3, we can bridge between a less-constrained mapping and an alignment between caterpillars as follows, see [6, 8] in more detail. We can construct a less-constrained mapping $M \in \mathcal{M}_{\text{LESS}}(T_1, T_2)$ from an alignment $\mathcal{T} \in \mathcal{A}(T_1, T_2)$ such that $(v, w) \in M$ iff $(l(v), l(w)) \in \mathcal{T}$. We denote such an M by $M_{\mathcal{T}}$. Conversely, we can construct an alignment $\mathcal{T} \in \mathcal{A}(T_1, T_2)$ from a less-constrained mapping $M \in \mathcal{M}_{\text{LESS}}(T_1, T_2)$ by setting $(l(v), l(w))$ if $(v, w) \in M$, $(l(v), \varepsilon)$ if $v \in M|_1$ and $(\varepsilon, l(w))$ if $w \in M|_2$ to labels of vertices in $\mathcal{T}$. We denote such a $\mathcal{T}$ by $\mathcal{T}_M$.

Lemma 1 *If $\mathcal{T} \in \mathcal{CA}(C_1, C_2)$, then $M_{\mathcal{T}} \in \mathcal{M}_{\text{CLESS}}(C_1, C_2)$.*

Proof By Theorem 3, it holds that $M_{\mathcal{T}} \in \mathcal{M}_{\text{LESS}}(C_1, C_2)$, so it is sufficient to show that $M_{\mathcal{T}}$ satisfies the condition in Definition 10.

Every vertex in $\mathcal{T}$ is one of the forms of $(l(v), l(w))$, $(l(v), \varepsilon)$ and $(\varepsilon, l(w))$, where $v \in C_1$ and $w \in C_2$. If $(l(v_1), l(w_1))$ such that $v_1 \in bb(C_1)$ and $w_1 \in lv(C_2)$ is a vertex in $\mathcal{T}$, then it is in $bb(\mathcal{T})$, because v_1 has descendants in C_1 and $(l(v_1), l(w_1))$ has descendants in $\mathcal{T}$, which is of the form of $(l(v_1'), \varepsilon)$ for every descendant v_1' of v_1 in C_1. In this case, there exists no vertex $(l(v_2), l(w_2))$ in $\mathcal{T}$ such that $v_2 \in lv(C_1)$ and

$w_2 \in bb(C_2)$. Because, if so, then $(l(v_2), l(w_2))$ has the descendants in $\mathcal{T}$, which is of the form of $(\varepsilon, l(w_2'))$ for every descendant w_2' of w_2 in C_2. Since $(l(v_1'), \varepsilon)$ is the descendants of $(l(v_1), l(w_1))$ in $\mathcal{T}$ and $(\varepsilon, l(w_2'))$ is the descendants of $(l(v_2), l(w_2))$ in $\mathcal{T}$, $\mathcal{T}$ has the vertex whose children are $(l(v_1), l(w_1))$ and $(l(v_2), l(w_2))$, which implies that $\mathcal{T}$ is not a caterpillar.

As a result, if $(l(v_1), l(w_1))$ such that $v_1 \in bb(C_1)$ and $w_1 \in lv(C_2)$ is a vertex in $\mathcal{T}$, then there exists no vertex $(l(v_2), l(w_2))$ in $\mathcal{T}$ such that $v_2 \in lv(C_1)$ and $w_2 \in bb(C_2)$. By using the same discussion, if $(l(v_2), l(w_2))$ such that $v_2 \in lv(C_1)$ and $w_2 \in bb(C_2)$ is a vertex in $\mathcal{T}$, then there exists no vertex $(l(v_1), l(w_1))$ in $\mathcal{T}$ such that $v_1 \in bb(C_1)$ and $w_1 \in lv(C_2)$. Hence, $M_{\mathcal{T}}$ has no two pairs (v_1, w_1) and (v_2, w_2) such that $v_1 \in bb(C_1)$, $w_1 \in lv(C_2)$, $v_2 \in lv(C_1)$ and $w_2 \in bb(C_2)$, which implies that $M_{\mathcal{T}} \in \mathcal{M}_{\text{CLESS}}(C_1, C_2)$. □

Lemma 2 *If M is the minimum cost caterpillar less-constrained mapping between C_1 and C_2, then $\mathcal{T}_M$ is a caterpillar.*

Proof By Definition 10, it holds that $M \in \mathcal{M}_{\text{LESS}}(C_1, C_2)$, so $\mathcal{T}_M$ is a tree by Theorem 3. Hence, it is sufficient to show that $\mathcal{T}_M$ is a caterpillar.

By Definition 5, $\mathcal{T}_M$ is obtained by merging two caterpillars constructed from C_1 and C_2 by inserting new vertex labeled by ε. Then, after removing all the leaves from $\mathcal{T}_M$, we obtain the tree $\mathcal{B}_M$ with one of the following $\mathcal{B}_1$ and $\mathcal{B}_2$:

$$\begin{aligned}
\mathcal{B}_1 &= (\{b_1, \ldots, b_l\}, \{(b_i, b_{i+1}) \mid 1 \leq i \leq l-1\}),\\
\mathcal{B}_2 &= (V(\mathcal{B}_1) \cup V(\mathcal{B}') \cup V(\mathcal{B}''), E(\mathcal{B}_1) \cup E(\mathcal{B}') \cup E(\mathcal{B}'') \cup \{(b_l, c_1), (b_l, d_1)\}),
\end{aligned}$$

where

$$\begin{aligned}
\mathcal{B}' &= (\{c_1, \ldots, c_m\}, \{(c_j, c_{j+1}) \mid 1 \leq j \leq m-1\}),\\
\mathcal{B}'' &= (\{d_1, \ldots, d_n\}, \{(d_k, d_{k+1}) \mid 1 \leq k \leq n-1\}).
\end{aligned}$$

If it holds that $\mathcal{B}_M = \mathcal{B}_1$, then it is obvious that $\mathcal{T}_M$ is a caterpillar.

Suppose that $\mathcal{B}_M = \mathcal{B}_2$. Let $c_j = (p_j, q_j)$ in $\mathcal{B}'$ $(1 \leq j \leq m-1)$ such that p_j is either ε or $l(v)$ for $v \in bb(C_1)$ and $d_k = (r_k, s_k)$ in $\mathcal{B}''$ $(1 \leq k \leq n-1)$ such that s_k is either ε or $l(w)$ for $w \in bb(C_2)$.

If $p_j = \varepsilon$, then $q_j = l(w)$ for $w \in lv(C_2)$. In this case, without loss of generality, we can place (p_j, q_j) in $lv(\mathcal{T}_M)$, not in $\mathcal{B}_2$. Hence, we can assume that every p_j is $l(v)$ for $v \in bb(C_1)$. By using the same discussion, we can assume that every s_k is also $l(w)$ for $w \in bb(C_2)$.

Every q_j is either ε or $l(w)$ for $w \in lv(C_2)$, because, if q_j is $l(w)$ for $w \in bb(C_2)$, then w is incomparable with w' for $s_k = l(w')$, which implies that M does not satisfy the ancestor condition. Also, there exists at most one q_j such that $q_j = l(w)$ for $w \in lv(C_2)$, because w has no descendants in C_2. Since $(p_j, q_j) \neq (\varepsilon, \varepsilon)$, it holds that (1) every (p_j, q_j) is $(l(v), \varepsilon)$ or $(l(v), l(w))$ for $v \in bb(C_1)$ and $w \in lv(C_2)$ and (2) $(l(v), l(w))$ occurs at most once. By using the same discussion, it holds that (3) every (r_k, s_k) is $(\varepsilon, l(w))$ or $(l(v), l(w))$ for $v \in lv(C_1)$ and $w \in bb(C_2)$ and (4) $(l(v), l(w))$ occurs at most once.

Suppose that every q_j is ε and every r_k is ε. Then, since $\gamma((p_j, \varepsilon)) + \gamma((\varepsilon, s_k)) \geq \gamma((p_j, s_k))$ for any cost function γ and M is the minimum cost, we can merge $\mathcal{B}'$ with $\mathcal{B}''$ as possible which has a smaller cost than $\gamma(\mathcal{B}' \cup \mathcal{B}'')$. In this case, we can regard $\mathcal{B}_2$ as $\mathcal{B}_1$.

Suppose that $q_j = l(w_1)$ for $w_1 \in lv(C_2)$ for some j and every r_k is ε. That is, suppose that $(p_j, q_j) = (l(v_1), l(w_1))$ and $(r_k, s_k) = (\varepsilon, l(w_2))$ for $v_1 \in bb(C_1)$, $w_1 \in lv(C_2)$ and $w_2 \in bb(C_2)$. By using the above discussion, we can assume that $j = 1$, that is, $(p_1, q_1) = (l(v_1), l(w_1))$ and $(p_j, q_j) = (l(v), \varepsilon)$ $(2 \leq j \leq m-1)$ for $v_1, v \in bb(C_1)$ and $w_1 \in lv(C_2)$. In this case, we can merge $\mathcal{B}' \setminus \{(p_1, q_1)\}$ with $\mathcal{B}''$ by starting (p_1, q_1) which has a smaller cost than $\gamma(\mathcal{B}' \cup \mathcal{B}'')$. Then, we can regard $\mathcal{B}_2$ as $\mathcal{B}_1$. When $r_k = l(v_1)$ for $v_1 \in lv(C_1)$ for some k and every q_j is ε, we obtain the same result.

Suppose that, for some j and k, $q_j = l(w_1)$ for $w_1 \in lv(C_2)$ and $r_k = l(v_2)$ for $v_2 \in lv(C_1)$. In this case, since $p_j = l(v_1)$ for $v_1 \in bb(C_1)$ and $s_k = l(w_2)$ for $w_2 \in bb(C_2)$, it holds that $(v_1, w_1), (v_2, w_2) \in M$, $v_1 \in bb(C_1)$, $v_2 \in lv(C_1)$, $w_1 \in lv(C_2)$ and $w_2 \in bb(C_2)$, which is a contradiction that $M \in \mathcal{M}_{\mathrm{CLESS}}(C_1, C_2)$.

Hence, $\mathcal{B}_M$ is always $\mathcal{B}_1$, which implies that $\mathcal{T}_M$ is a caterpillar.

Theorem 4 *For caterpillars C_1 and C_2, it holds that $\tau_{\mathrm{CALN}}(C_1, C_2) = \tau_{\mathrm{CLESS}}(C_1, C_2)$.*

Proof By Lemma 1, it holds that $\gamma(\mathcal{T}) = \gamma(M_{\mathcal{T}})$ for $\mathcal{T} \in \mathcal{CA}(C_1, C_2)$. Since $\{M_{\mathcal{T}} \mid \mathcal{T} \in \mathcal{CA}(C_1, C_2)\} \subseteq \{M \mid M \in \mathcal{M}_{\mathrm{CLESS}}(C_1, C_2)\}$, it holds that $\tau_{\mathrm{CALN}}(C_1, C_2) = \min \{\gamma(\mathcal{T}) \mid \mathcal{T} \in \mathcal{CA}(C_1, C_2)\} = \min\{\gamma(M_{\mathcal{T}}) \mid \mathcal{T} \in \mathcal{CA}(C_1, C_2)\} \geq \min\{\gamma(M) \mid M \in \mathcal{M}_{\mathrm{CLESS}}(C_1, C_2)\} = \tau_{\mathrm{CLESS}}(C_1, C_2)$.

On the other hand, we denote the set of the minimum cost caterpillar less-constrained mappings between C_1 and C_2 by $\mathcal{M}^*_{\mathrm{CLESS}}(C_1, C_2)$ and let $M^* \in \mathcal{M}^*_{\mathrm{CLESS}}$ (C_1, C_2). Then, it is obvious that $\tau_{\mathrm{CLESS}}(C_1, C_2) = \gamma(M^*)$. By Lemma 2, $\mathcal{T}_{M^*}$ is a caterpillar, so it holds that $\gamma(\mathcal{T}_{M^*}) \geq \min\{\gamma(\mathcal{T}) \mid \mathcal{T} \in \mathcal{CA}(C_1, C_2)\} = \tau_{\mathrm{CALN}}(C_1, C_2)$. Hence, it holds that $\tau_{\mathrm{CLASS}}(C_1, C_2) \geq \tau_{\mathrm{CALN}}(C_1, C_2)$. □

Theorem 5 presents the relationship between τ_{CALN} and the variations of τ_{TAI}.

Theorem 5 *For caterpillars C_1 and C_2, it holds that $\tau_{\mathrm{ALN}}(C_1, C_2) \leq \tau_{\mathrm{CALN}}(C_1, C_2)$ but the equality does not always hold in general. On the other hand, $\tau_{\mathrm{CALN}}(C_1, C_2)$ and $\tau_{\mathrm{ILST}}(C_1, C_2)$ are incomparable.*

Proof The first statement is obvious by Definition 8, Examples 4 and 5.

By Example 2 and 6, the caterpillars C_1 and C_2 in Fig. 2 satisfy that $3 = \tau_{\mathrm{CLESS}}(C_1, C_2) = \tau_{\mathrm{CALN}}(C_1, C_2) < \tau_{\mathrm{ILST}}(C_1, C_2) = 5$. On the other hand, by Example 4 and 5 and since $M_5 \in \mathcal{M}_{\mathrm{ILST}}(C_5, C_6)$, the caterpillars C_5 and C_6 in Fig. 5 satisfy that $2 = \tau_{\mathrm{ILST}}(C_5, C_6) < \tau_{\mathrm{CALN}}(C_5, C_6) = 3$. Hence, the second statement holds. □

Thorem 5 implies that τ_{CALN} and not only τ_{ILST} but also other variations of τ_{TAI} defined by more restricted mappings than $\mathcal{M}_{\mathrm{ILST}}$ (*cf.*, [16]) such as an LCA-preserving (or degree-2) distance [19] and a top-down distance [2, 12] are incomparable.

Finally, the following example claims that the caterpillar alignment distance τ_{CALN} is not a metric as same as the alignment distance τ_{ALN}.

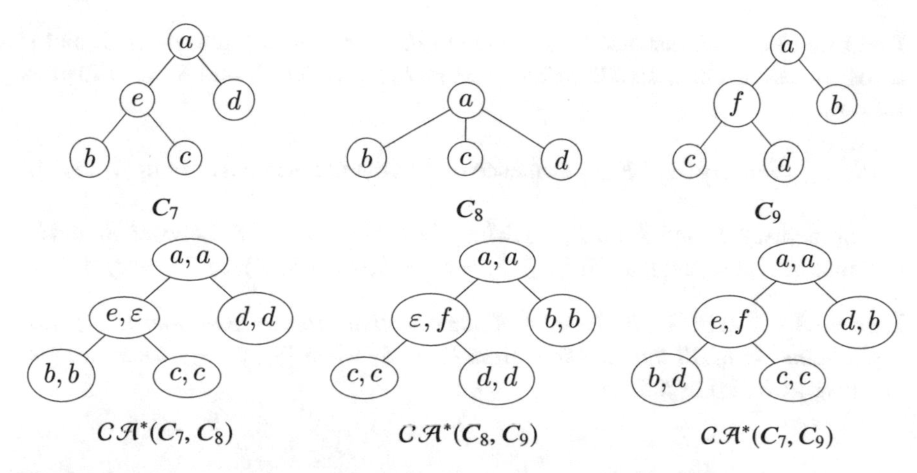

Fig. 7 Caterpillars C_7, C_8 and C_9 and the optimal caterpillar alignments $\mathcal{CA}^*(C_7, C_8)$, $\mathcal{CA}^*(C_8, C_9)$ and $\mathcal{CA}^*(C_7, C_9)$ in Example 7

Example 7 Consider the caterpillars C_7, C_8 and C_9 illustrated in Fig. 7, which are same as trees in [8]. Then, we obtain the optimal caterpillar alignments $\mathcal{CA}^*(C_7, C_8)$, $\mathcal{CA}^*(C_8, C_9)$ and $\mathcal{CA}^*(C_7, C_9)$ illustrated in Fig. 7. Hence, it holds that $\tau_{\text{CALN}}(C_7, C_8) + \tau_{\text{CALN}}(C_8, C_9) = 1 + 1 = 2 < 3 = \tau_{\text{CALN}}(C_7, C_9)$, which implies that τ_{CALN} is not a metric.

5 The Algorithm of Computing Caterpillar Alignment Distance

In this section, we design the algorithm to compute the caterpillar alignment distance τ_{CALN} and analyze the time complexity of computing τ_{CALN}.

In order to compute the edit distance between the sets of leaves, it is necessary to introduce an edit distance for *multisets* on labels occurring in the set of leaves. Then, we prepare the notions of the edit distance for multisets according to [11].

A *multiset* on an alphabet Σ is a mapping $S : \Sigma \to \mathbf{N}$. For a multiset S on Σ, we say that $a \in \Sigma$ is an *element* of S if $S(a) > 0$ and denote it by $a \in S$ (like as a standard set). The *cardinality* of S, denoted by $|S|$, is defined as $\sum_{a \in \Sigma} S(a)$.

Definition 11 (Edit operations for multisets) Let $a, b \in \Sigma$ such that $S(a) > 0$ and $a \neq b$. Then, a *substitution* $(a \mapsto b)$ operates $S(a)$ to $S(a) - 1$ and $S(b)$ to $S(b) + 1$, a *deletion* $(a \mapsto \varepsilon)$ operates $S(a)$ to $S(a) - 1$ and an *insertion* $(\varepsilon \mapsto b)$ operates $S(b)$ to $S(b) + 1$.

Also we assume a cost function γ as in Sect. 2.

Definition 12 (Edit distance for multisets) Let S_1 and S_2 be multisets on Σ and γ a cost function. Then, an *edit distance* $\mu(S_1, S_2)$ between S_1 and S_2 is defined as follows.

$$\mu(S_1, S_2) = \min\{\gamma(E) \mid E \text{ is a sequence of edit operations transforming } S_1 \text{ to } S_2\}.$$

For multisets S_1 and S_2 on Σ, we define the *difference* $S_1 \setminus S_2$ between S_1 and S_2 as a multiset satisfying that $(S_1 \setminus S_2)(a) = \max\{S_1(a) - S_2(a), 0\}$ for every $a \in \Sigma$.

Lemma 3 ([11]) *Let* Π_1 *be the set of all the injections from* S_1 *to* S_2 *when* $|S_1| \leq |S_2|$ *and* Π_2 *the set of all the injections from* S_2 *to* S_1 *when* $|S_1| > |S_2|$. *Then, we can compute* $\mu(S_1, S_2)$ *as follows:*

$$\mu(S_1, S_2) = \begin{cases} \min\limits_{\pi \in \Pi_1} \left\{ \sum\limits_{a \in S_1} \gamma(a, \pi(a)) + \sum\limits_{b \in S_2 \setminus \pi(S_1)} \gamma(\varepsilon, b) \right\}, & \textit{if } |S_1| \leq |S_2|, \\ \min\limits_{\pi \in \Pi_2} \left\{ \sum\limits_{b \in S_2} \gamma(\pi(b), b) + \sum\limits_{a \in S_1 \setminus \pi(S_2)} \gamma(a, \varepsilon) \right\}, & \textit{otherwise.} \end{cases}$$

Furthermore, if we adopt the unit cost function, then we can compute $\mu(S_1, S_2)$ *as follows:*

$$\mu(S_1, S_2) = \max\{|S_1 \setminus S_2|, |S_2 \setminus S_1|\}.$$

In this case, $\mu(S_1, S_2)$ *coincides with a famous* bag distance *(cf., [3]) between* S_1 *and* S_2.

Lemma 4 ([11]) *Let* $m = \max\{|S_1|, |S_2|\}$. *Then, we can compute* $\mu(S_1, S_2)$ *in* $O(m^3)$ *time under the general cost function. If we adopt the unit cost function, then we can compute* $\mu(S_1, S_2)$ *in* $O(m)$ *time.*

Next, we design the recurrences of computing τ_{CALN}.

Let L be the set of leaves and C a non-leaf caterpillar. Then, every forest obtained by deleting the root from a caterpillar is one of the forms of $\{C\}$, L or $L \cup \{C\}$. As same as [10], we denote these forests by $\langle \emptyset \mid C \rangle$, $\langle L \mid \emptyset \rangle$ and $\langle L \mid C \rangle$, respectively. In particular, we denote an empty forest $\langle \emptyset \mid \emptyset \rangle$ by Φ simply.

Let $C[v]$ be a caterpillar with the root v, where $L(v)$ denotes a (possibly empty) set of leaves as the children of v and $B(v)$ denotes at most one caterpillar of the child v. Then, $C[v]$ is one of the forms in Fig. 8. Furthermore, by deleting v from $C[v]$, we obtain one of the forests of $\langle \emptyset \mid B(v) \rangle$, $\langle L(v) \mid \emptyset \rangle$ and $\langle L(v) \mid B(v) \rangle$, respectively.

Figure 9 illustrates the recurrences of computing $\tau_{\text{CALN}}(C_1[v], C_2[w])$ between two caterpillars $C_1[v]$ and $C_2[w]$. Here, we regard a set L of leaves as a multiset of labels on Σ occurring in L, which we denote by $\widetilde{L}$. Also $\delta_{\text{CALN}}(\langle L_1 \mid B_1 \rangle, \langle L_2 \mid B_2 \rangle)$ describes the caterpillar alignment distance between forests $\langle L_1 \mid B_1 \rangle$ and $\langle L_2 \mid B_2 \rangle$. Furthermore, assume that the forest obtained by deleting v (*resp.*, w) from $C_1[v]$

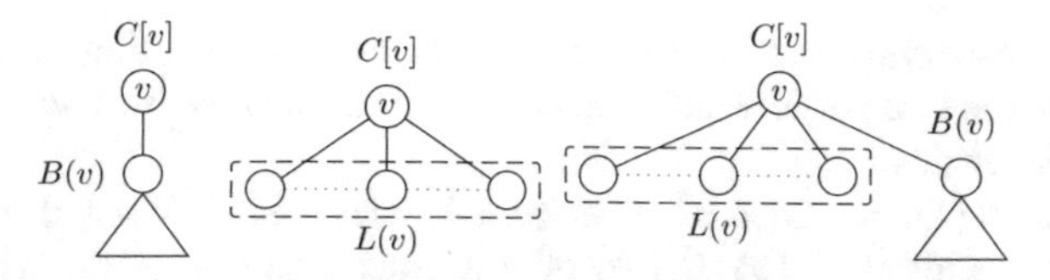

Fig. 8 The representation of a caterpillar $C[v]$

$$\tau_{\text{CALN}}(\emptyset, \emptyset) = 0. \quad (T_0)$$
$$\tau_{\text{CALN}}(C_1[v], \emptyset) = \gamma(v, \varepsilon) + \delta_{\text{CALN}}(\langle L_1(v) \mid B_1(v)\rangle, \Phi). \quad (T_1)$$
$$\tau_{\text{CALN}}(\emptyset, C_2[w]) = \gamma(\varepsilon, w) + \delta_{\text{CALN}}(\Phi, \langle L_2(w) \mid B_2(w)\rangle). \quad (T_2)$$
$$\delta_{\text{CALN}}(\langle L_1 \mid C_1\rangle, \Phi) = \sum_{v \in L_1} \gamma(v, \varepsilon) + \sum_{v \in C_1} \gamma(v, \varepsilon). \quad (F_1)$$
$$\delta_{\text{CALN}}(\Phi, \langle L_2 \mid C_2\rangle) = \sum_{w \in L_2} \gamma(\varepsilon, w) + \sum_{w \in C_2} \gamma(\varepsilon, w). \quad (F_2)$$
$$\delta_{\text{CALN}}(\langle L_1 \mid \emptyset\rangle, \langle L_2 \mid \emptyset\rangle) = \mu(\widetilde{L_1}, \widetilde{L_2}). \quad (F_3)$$

(A) $\tau_{\text{CALN}}(C_1[v], C_2[w])$

$$= \min \begin{Bmatrix} \gamma(v, w) + \delta_{\text{CALN}}(\langle L_1(v) \mid B_1(v)\rangle, \langle L_2(w) \mid B_2(w)\rangle), & (T_3) \\ \gamma(v, w) + \delta_{\text{CALN}}(\langle \emptyset \mid C_1[v := \varepsilon]\rangle, \langle L_2(w) \mid B_2(w)\rangle), & (T_4) \\ \gamma(v, w) + \delta_{\text{CALN}}(\langle L_1(v) \mid B_1(v)\rangle, \langle \emptyset \mid C_2[w := \varepsilon]\rangle), & (T_5) \\ \gamma(v, \varepsilon) + \tau_{\text{CALN}}(B_1(v), C_2[w]) + \delta_{\text{CALN}}(\langle L_1(v) \mid \emptyset\rangle, \Phi), & (T_6) \\ \gamma(\varepsilon, w) + \tau_{\text{CALN}}(C_1[v], B_2(w)) + \delta_{\text{CALN}}(\Phi, \langle L_2(w) \mid \emptyset\rangle) & (T_7) \end{Bmatrix}.$$

(B) $\delta_{\text{CALN}}(\langle L_1 \mid \emptyset\rangle, \langle L_2 \mid C_2[w]\rangle)$

$$= \min \begin{Bmatrix} \gamma(\varepsilon, w) + \delta_{\text{CALN}}(\langle L_1 \mid \emptyset\rangle, \langle L_2 \cup L_2(w) \mid B_2(w)\rangle), & (F_4) \\ \min_{v \in L_1} \{\gamma(v, w) + \delta_{\text{CALN}}(\langle L_1 \setminus \{v\} \mid \emptyset\rangle, \langle L_2 \mid \emptyset\rangle)\} + \delta_{\text{CALN}}(\Phi, \langle L_2(w) \mid B_2(w)\rangle) & (F_5) \end{Bmatrix}.$$

(C) $\delta_{\text{CALN}}(\langle L_1 \mid C_1[v]\rangle, \langle L_2 \mid \emptyset\rangle)$

$$= \min \begin{Bmatrix} \gamma(v, \varepsilon) + \delta_{\text{CALN}}(\langle L_1 \cup L_1(v) \mid B_1(v)\rangle, \langle L_2 \mid \emptyset\rangle), & (F_6) \\ \min_{w \in L_2} \{\gamma(v, w) + \delta_{\text{CALN}}(\langle L_1 \mid \emptyset\rangle, \langle L_2 \setminus \{w\} \mid \emptyset\rangle)\} + \delta_{\text{CALN}}(\langle L_1(v) \mid B_1(v)\rangle, \Phi) & (F_7) \end{Bmatrix}.$$

(D) $\delta_{\text{CALN}}(\langle L_1 \mid C_1[v]\rangle, \langle L_2 \mid C_2[w]\rangle) = \delta_{\text{CALN}}(\langle L_1 \mid \emptyset\rangle, \langle L_2 \mid \emptyset\rangle) + \tau_{\text{CALN}}(C_1[v], C_2[w])$. (F_8)

Fig. 9 The recurrences of computing $\tau_{\text{CALN}}(C_1[v], C_2[w])$ between $C_1[v]$ and $C_2[w]$

(*resp.*, $C_2[w]$) is $\langle L_1(v) \mid B_1(v)\rangle$ (*resp.*, $\langle L_2(w) \mid B_2(w)\rangle$). Finally, for caterpillars $C_1[v]$ and $C_2[w]$, $C_1[v := \varepsilon]$ and $C_2[w := \varepsilon]$ denote the caterpillars obtained by replacing v in C_1 and w in C_2 with the vertex labeled by ε.

Theorem 6 *The recurrences in Fig. 9 are correct to compute the caterpillar alignment distance $\tau_{\text{CALN}}(C_1[v], C_2[w])$ between $C_1[v]$ and $C_2[w]$.*

Proof The recurrences of (T_0), (T_1), (T_2), (F_1), (F_2) and (F_3) are obvious.

First, consider the recurrences for τ_{CALN}. Let $\mathcal{T}$ be the optimal caterpillar alignment (caterpillar) $\mathcal{A}^*(C_1[v], C_2[w])$. Then, for the label in $\mathcal{T}$, one of the following four cases holds.

1. (v, w) is a label in $\mathcal{T}$.
2. (v, ε) and (v', w) are labels in $\mathcal{T}$.
3. (ε, w) and (v, w') are labels in $\mathcal{T}$.
4. (v, ε) and (ε, w) are labels in $\mathcal{T}$.

It is not necessary to consider the case (4) because the resulting caterpillar alignment to delete the two vertices and then add (v, w) as the new root, which is the case (1), has a smaller cost.

For the case (1), the root of $\mathcal{T}$ is (v, w). Then, there exist three cases that (1) the whole children $\langle L_1(v) \mid B_1(v)\rangle$ of v is corresponding to the whole children $\langle L_2(w) \mid B_2(w)\rangle$ of w, (2) the caterpillar $B_2(w)$ is corresponding to the whole children $\langle L_1(v) \mid B_1(v)\rangle$ of v and (c) the caterpillar $B_1(v)$ is corresponding to the whole children $\langle L_2(w) \mid B_2(w)\rangle$ of w. For the case (a), by the forms of $C_1[v]$ and $C_2[w]$, it holds that $\tau_{\text{CALN}}(C_1[v], C_2[w]) = \gamma(v, w) + \delta_{\text{CALN}}(\langle L_1(v) \mid B_1(v)\rangle, \langle L_2(w) \mid B_2(w)\rangle)$, which is the recurrence (T_3). For the case (b), by inserting a dummy vertex labeled by ε to the child of v, we regard the whole children of v as a caterpillar. Then, it holds that $\tau_{\text{CALN}}(C_1[v], C_2[w]) = \gamma(v, w) + \delta_{\text{CALN}}(\langle \emptyset \mid C_1[v := \varepsilon]\rangle, \langle L_2(w) \mid B_2(w)\rangle)$, which is the recurrence (T_4). As same as the case (b), for case (c), it holds that $\tau_{\text{CALN}}(C_1[v], C_2[w]) = \gamma(v, w) + \delta_{\text{CALN}}(\langle L_1(v) \mid B_1(v)\rangle, \langle \emptyset \mid C_2[w := \varepsilon]\rangle)$, which is the recurrence (T_5).

For the case (2), the root of $\mathcal{T}$ is (v, ε). By the form of $C_1[v]$, since $|B_1(v)| \geq 2$, $B_1(v)$, not $L_1(v)$, contains the vertex v' corresponding to w in $C_2[w]$. Then, $\mathcal{T}$ contains a label (v'', ε) for every $v'' \in L_1(v)$. Hence, it holds that $\tau_{\text{CALN}}(C_1[v], C_2[w]) = \gamma(v, \varepsilon) + \tau_{\text{CALN}}(B_1(v), C_2[w]) + \delta_{\text{CALN}}(\langle L_1(v) \mid \emptyset\rangle, \Phi)$, which is the recurrence (T_6). The case (3) is similar to the case (2), which is the recurrence (T_7).

Next, consider the recurrences for δ_{CALN}.

Let $\mathcal{F}$ be the optimal caterpillar alignment (forest) $\mathcal{A}^*(\langle L_1 \mid \emptyset\rangle, \langle L_2 \mid C_2[w]\rangle)$. Then, for the label in $\mathcal{F}$, one of the following two cases holds.

1. (ε, w) is a label in $\mathcal{F}$.
2. (v, w) for some $v \in L_1$ is a label in $\mathcal{F}$.

For the case (1), by deleting w from $C_2[w]$, $\langle L_2 \mid C_2[w]\rangle$ is transformed to $\langle L_2 \cup L_2(w) \mid B_2(w)\rangle$. Hence, it holds that $\delta_{\text{CALN}}(\langle L_1 \mid \emptyset\rangle, \langle L_2 \mid C_2[w]\rangle) = \gamma(\varepsilon, w) + \delta_{\text{CALN}}(\langle L_1 \mid \emptyset\rangle, \langle L_2 \cup L_2(w) \mid B_2(w)\rangle)$, which is the recurrence (F_4).

For the case (2), once (v, w) for some $v \in L_1$ becomes a label in $\mathcal{F}$, every label in $\mathcal{F}$ for every $w' \in \langle L_2(w) \mid B_2(w)\rangle$ is always of the form (ε, w'). Also the labels concerned with leaves except $v \in L_1$ in $\mathcal{F}$ can be computed as $\delta_{\text{CALN}}(\langle L_1 \setminus \{v\} \mid \emptyset\rangle, \langle L_2 \mid \emptyset\rangle)$. Hence, by selecting $v \in L_1$ with the minimum cost, we obtain the recurrence (F_5).

By using the same discussion, for the case that $\mathcal{F}$ is the optimal caterpillar alignment (forest) $\mathcal{A}^*(\langle L_1 \mid C_1[v]\rangle, \langle L_2 \mid \emptyset\rangle)$, we obtain the recurrences (F_6) and (F_7).

Let $\mathcal{F}$ be the optimal caterpillar alignment (forest) $\mathcal{A}^*(\langle L_1 \mid C_1[v]\rangle, \langle L_2 \mid C_2[w]\rangle)$. Since $\|C_1[v]\| \geq 2$ and $\|C_2[w]\| \geq 2$, $\mathcal{F}$ contains labels for the caterpillar alignment of $C_1[v]$ and $C_2[w]$ and that of L_1 and L_2. Hence, it holds that $\delta_{\text{CALN}}(\langle L_1 \mid C_1[v]\rangle, \langle L_2 \mid C_2[w]\rangle) = \delta_{\text{CALN}}(\langle L_1 \mid \emptyset\rangle, \langle L_2 \mid \emptyset\rangle + \tau_{\text{CALN}}(C_1[v], C_2[w])$, which is the recurrence (F_8). □

Example 8 Consider two caterpillars C_1 and C_2 in Fig. 2 in Example 1 and assume the unit cost function. By applying the recurrences in Fig. 9, we obtain that the caterpillar alignment distance $\tau_{\text{CALN}}(C_1, C_2)$ between C_1 and C_2 is 3 illustrated in

$$
\begin{aligned}
&\tau_{\text{CALN}}(C_1, C_2)\\
&= \underbrace{\gamma(a, a)}_{=0} + \delta_{\text{CALN}}(\langle [b, b, b] \mid \emptyset \rangle, \langle [b] \mid a[b, b[a, a]] \rangle) && (T_3)\\
&= \underbrace{\gamma(\varepsilon, a)}_{=1} + \delta_{\text{CALN}}(\langle [b, b, b] \mid \emptyset \rangle, \langle [b, b] \mid b[a, a] \rangle) && (F_4)\\
&= 1 + \underbrace{\gamma(b, b)}_{=0} + \delta_{\text{CALN}}(\langle [b, b] \mid \emptyset \rangle, \langle [b, b] \mid \emptyset \rangle) + \delta_{\text{CALN}}(\Phi, \langle [a, a] \mid \emptyset \} \rangle) && (F_5)\\
&= 1 + \underbrace{\mu([b, b], [b, b])}_{=0} + \underbrace{\gamma(\varepsilon, a)}_{=1} + \underbrace{\gamma(\varepsilon, a)}_{=1} && (F_2), (F_3)\\
&= 3.
\end{aligned}
$$

Fig. 10 The result of computing $\tau_{\text{CALN}}(C_1, C_2)$ in Example 8

$$
\begin{aligned}
&\tau_{\text{CALN}}(C_3, C_4)\\
&= \underbrace{\gamma(b, b)}_{0} + \delta_{\text{CALN}}(\langle [a] \mid a[b, c] \rangle, \langle [c] \mid c[b, a] \rangle) && (T_3)\\
&= \underbrace{\gamma(a, c)}_{1} + \delta_{\text{CALN}}(\langle [a] \mid \emptyset \rangle, \langle [c] \mid \emptyset \rangle) + \delta_{\text{CALN}}(\langle [b, c] \mid \emptyset \rangle, \langle [b, a] \mid \emptyset \rangle) && (F_8)\\
&= 1 + \underbrace{\mu([a], [c])}_{=1} + \underbrace{\mu([b, c], [a, b])}_{=1} && (F_3)\\
&= 3.
\end{aligned}
$$

Fig. 11 The result of computing $\tau_{\text{CALN}}(C_3, C_4)$ in Example 9

Fig. 10. Here, we represent a multiset as a sequence enclosed by "[" and "]" and a caterpillar as a term-like representation with "[" and "]", that is, $C_1 = a[b, b, b]$ and $C_2 = a[b, a[b, b[a, a]]]$.

Example 9 Consider two caterpillars $C_3 = a[a, d[b, c]]$ and $C_4 = a[c, e[b, a]]$ in Fig. 2 in Example 1 and assume the unit cost function. By applying the recurrences in Fig. 9, we obtain that the caterpillar alignment distance $\tau_{\text{CALN}}(C_3, C_4)$ between C_3 and C_4 is 3 illustrated in Fig. 11.

Example 10 In order to show the necessity of (T_4) and (T_5), consider two caterpillars $C_{10} = a[b[c, b]]$ and $C_{11} = a[c, b[e]]$ in Fig. 12 and assume the unit cost function. By applying the recurrences in Fig. 9, we obtain that the caterpillar alignment distance $\tau_{\text{CALN}}(C_{10}, C_{11})$ between C_{10} and C_{11} is 2 illustrated in Fig. 13.

Note that the special caterpillar $\varepsilon[c, b[d]]$ in the second formula in Fig. 13 is obtained by applying (T_5) in Fig. 9. The resulting caterpillar alignment $\mathcal{CA}^*(C_{10}, C_{11})$ and the resulting caterpillar less-constrained mapping M_{10} are illustrated in Fig. 9.

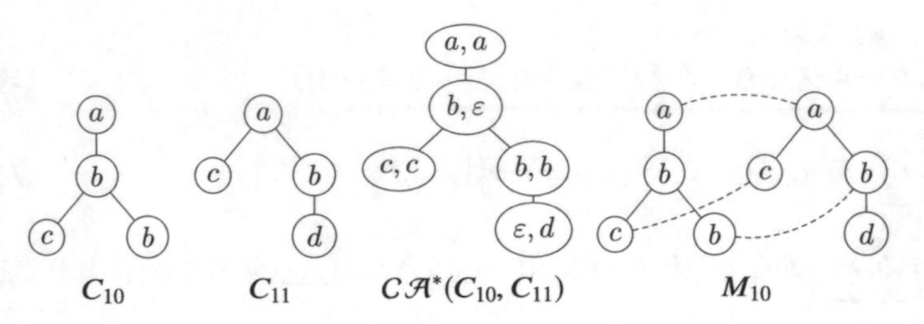

Fig. 12 Caterpillars C_{10} and C_{11}, the optimal caterpillar alignment $C\mathcal{A}^*(C_{10}, C_{11})$ and the minimum cost caterpillar less-constrained mapping $M_{10} \in \mathcal{M}_{\text{CLESS}}(C_{10}, C_{11})$ in Example 10

$$
\begin{aligned}
&\tau_{\text{CALN}}(C_{10}, C_{11}) \\
&= \underbrace{\gamma(a, a)}_{0} + \delta_{\text{CALN}}(\langle \emptyset \,|\, b[c, b] \rangle, \langle \emptyset \,|\, \varepsilon[c, b[d]] \rangle) && (T_5) \\
&= \tau_{\text{CALN}}(b[c, b], \varepsilon[c, b[d]]) && (F_8) \\
&= \underbrace{\gamma(b, \varepsilon)}_{1} + \delta_{\text{CALN}}(\langle [c, b] \,|\, \emptyset \rangle, \langle [c] \,|\, b[d] \rangle) && (T_3) \\
&= 1 + \underbrace{\gamma(b, b)}_{=0} + \delta_{\text{CALN}}(\langle [c] \,|\, \emptyset \rangle, \langle [c] \,|\, \emptyset \rangle) + \delta_{\text{CALN}}(\Phi, \langle [d] \,|\, \emptyset \rangle) && (F_5) \\
&= 1 + \underbrace{\mu([c], [c])}_{=0} + \underbrace{\gamma(d, \varepsilon)}_{=1} && (F_2), (F_3) \\
&= 2.
\end{aligned}
$$

Fig. 13 The result of computing $\tau_{\text{CALN}}(C_{10}, C_{11})$ in Example 10

Finally, by using the recurrences in Fig. 9, we design the algorithm of computing τ_{CALN} and then analyze its time complexity.

Let $C_1[v]$ and $C_2[w]$ be caterpillars. Then, we denote $bb(C_1[v])$ by a sequence $v_1, \ldots, v_n$ such that $v_n = v$ and $par(v_i) = v_{i+1}$ $(1 \leq i \leq n-1)$ and $bb(C_2[w])$ by a sequence $w_1, \ldots, w_m$ such that $w_m = w$ and $par(w_j) = w_{j+1}$ $(1 \leq j \leq m-1)$. In this case, we denote by $bb(C_1[v]) = [v_1, \ldots, v_n]$ and $bb(C_2[w]) = [w_1, \ldots, w_m]$. Also we use the same notations of $L_1(v_i)$ and $B_1(v_i)$ for $1 \leq i \leq n$ and $L_2(w_j)$ and $B_2(w_j)$ for $1 \leq j \leq m$.

Based on the recurrences in Fig. 9, Algorithm 1 illustrates the algorithm to compute the caterpillar alignment distance $\tau_{\text{CALN}}(C_1, C_2)$ between caterpillars C_1 and C_2. Here, the statement "$v \leftarrow$ (A)" means to substitute the value of computing the right side of the recurrence (A) to v, for example.

Theorem 7 *Let C_1 and C_2 be caterpillars. Then, we can compute the caterpillar alignment distance $\tau_{\text{CALN}}(C_1, C_2)$ between C_1 and C_2 in $O(h^2\lambda^3)$ time, where $h = \max\{h(C_1), h(C_2)\}$ and $\lambda = \max\{|lv(C_1)|, |lv(C_2)|\}$. Furthermore, if we adopt the unit cost function, then we can compute it in $O(h^2\lambda)$ time.*

Proof Let $bb(C_1) = [v_1, \ldots, v_n]$ and $bb(C_2) = [w_1, \ldots, w_m]$. Then, it is obvious that $h(C_1) = n + 1$ and $h(C_2) = m + 1$, so it holds that $m \leq h - 1$ and $n \leq h - 1$.

```
procedure τ_CALN(C1, C2)
  /* C1, C2: caterpillars, bb(C1) = [v1, ..., vn], bb(C2) = [w1, ..., wm], vn = r(C1),
  wm = r(C2) */ τ_CALN(∅, ∅) ← 0; /* (T0) */ for i = 1 to n do τ_CALN(C1[vi], ∅) ← (T1);
  for j = 1 to m do τ_CALN(∅, C2[wj]) ← (T2);
  for i = 1 to n do
    for j = 1 to m do
      τ_CALN(C1[vi], C2[wj]) ← (A);

procedure δ_CALN(⟨L1 | C1⟩, ⟨L2 | C2⟩)
  /* L1, L2 : set of leaves, C1, C2: caterpillars */ if C1 = ∅ and C2 = ∅ then
  δ_TAI(⟨L1 | ∅⟩, ⟨L2 | ∅⟩) ← (F3);
  else if C1 ≠ ∅ and C2 = ∅ then
    /* bb(C1) = [v1, ..., vn], vn = r(C1) */ if L2 = ∅ then δ_TAI(⟨L1 | C1⟩, Φ) ← (F1);
    else
      for i = 1 to n do δ_TAI(⟨L1 | C1[vi]⟩, ⟨L2 | ∅⟩) ← (B);
  else if C1 = ∅ and C2 ≠ ∅ then
    /* bb(C2) = [w1, ..., wm], wm = r(C2) */
    if L1 = ∅ then δ_TAI(Φ, ⟨L2 | C2⟩) ← (F2);
    else
      for j = 1 to m do δ_TAI(⟨L1 | ∅⟩, ⟨L2 | C2[wj]⟩) ← (C);
  else
    /* bb(C1) = [v1, ..., vn], bb(C2) = [w1, ..., wm], vn = r(C1), wm = r(C2) */
    for i = 1 to n do
      for j = 1 to m do
        δ_TAI(⟨L1 | C1[vi]⟩, ⟨L2 | C2[wj]⟩) ← (D);
```

Algorithm 1: $\tau_{\mathrm{CALN}}(C_1, C_2)$

The algorithm of computing $\tau_{\mathrm{CALN}}(C_1, C_2)$ calls every pair $(v_i, w_j) \in bb(C_1) \times bb(C_2)$ just once. When computing $\delta_{\mathrm{CALN}}(\langle L_1(v_{i-1}) \mid C_1[v_i]\rangle, \langle L_2(w_{j-1}) \mid C_2[w_j]\rangle)$ for $2 \leq i \leq n$ and $2 \leq j \leq m$, it is possible to construct multisets $S_1 = L_1(v_1) \sqcup \cdots \sqcup L_1(v_{i-1})$ and $S_2 = L_2(w_1) \sqcup \cdots \sqcup L_2(w_{j-1})$ and compute the edit distance $\mu(S_1, S_2)$ between multisets in the worst case. By Lemma 4, we can compute it in $O(\lambda^3)$ time under the general cost function and in $O(\lambda)$ time under the unit cost function.

Hence, the total running time of computing $\tau_{\mathrm{CALN}}(C_1, C_2)$ under the general cost function is described as follows:

$$\sum_{i=1}^{n}\sum_{j=1}^{m} O(\lambda^3) = O(\lambda^3)mn \leq O(\lambda^3)(h-1)^2 = O(h^2\lambda^3).$$

If we adopt the unit cost function, then, by Lemma 4, we can replace $O(\lambda^3)$ with $O(\lambda)$. Hence, the above time complexity is reduced to $O(h^2\lambda)$ time under the unit cost function. □

Finally, concerned with Theorem 4, we discuss the related hardness results of computing alignment distances.

We say that a tree is a *generalized caterpillar* if it is transformed to a caterpillar after removing all the leaves in it. Also we define the *generalized caterpillar alignment distance* τ_{GCALN} as the alignment distance whose alignment is a generalized caterpillar. Then, the following theorem holds as a corollary of [10].

Theorem 8 (*cf.*, [1, 5]) *Let C_1 and C_2 be generalized caterpillars. Then, the problems of computing $\tau_{\mathrm{ALN}}(C_1, C_2)$ and $\tau_{\mathrm{GCALN}}(C_1, C_2)$ are* MAX SNP*-hard, even if the maximum height is at most* 3.

Proof It is straightforward from the proof of Corollary 4.3 in [1] or Theorem 1 in [5] and because the Tai mapping and the alignment constructed in their proof are a less-constrained mapping and a generalized caterpillar, respectively. □

Note that, if C_1 and C_2 are generalized caterpillars but not caterpillars, then the alignment of C_1 and C_2 is not a caterpillar, because the alignment is constructed by inserting vertices (without removing vertices) to C_1 and C_2. Then, $\tau_{\mathrm{CALN}}(C_1, C_2)$ for generalized caterpillars C_1 and C_2 is meaningless. Hence, Theorem 4 and 8 imply that the structural restriction of caterpillars provides some kinds of limitation for tractable computing the alignment distance for unordered trees

On the other hand, it is still open whether or not the problem of computing $\tau_{\mathrm{ALN}}(C_1, C_2)$ for caterpillars C_1 and C_2 is tractable.

6 Experimental Results

In this section, we give experimental results for computing the caterpillar alignment distance τ_{CALN}. Throughout of this section, we assume that *the cost function is the unit cost function*.

First, we explain caterpillars in real data, which we will discuss later. Table 1 illustrates the number of caterpillars in N-glycans and all glycans from KEGG,[1] CSLOGS,[2] dblp,[3] and SwissProt, TPC-H, Auction, Nasa, Protein and University from UW XML Repository.[4] Here, #cat is the number of caterpillars and #data is the total number of data.

For $D \in$ {Auction, Nasa, Protein, University}, D^- denotes the trees obtained by deleting the root for every tree in D. Since one tree in D produces some trees in D^-, the total number of trees in D^- is greater than that of D. Hence, there are some cases containing many caterpillars in real dataset.

[1]Kyoto Encyclopedia of Genes and Genomes, http://www.kegg.jp/.

[2]http://www.cs.rpi.edu/~zaki/www-new/pmwiki.php/Software/Software.

[3]http://dblp.uni-trier.de/.

[4]http://aiweb.cs.washington.edu/research/projects/xmltk/xmldata/www/repository.html.

Table 1 The number of caterpillars in N-glycans and all glycans from KEGG, CSLOGS, dblp, SwissProt, TPC-H, Auction, University, Protein and NASA

dataset	#cat	#data	%
N-glycans	514	2,142	23.996
All glycans	7,984	10,704	74.785
CSLOGS	41,592	59,691	69.679
dblp	5,154,295	5,154,530	99.995
SwissProt	6,804	50,000	13.608
TPC-H	86,805	86,805	100.000
Auction	0	37	0
Nasa	0	2,430	0
Protein	0	262,625	0
University	0	6,738	0
Auction$^-$	259	259	100.000
Nasa$^-$	21,245	27,921	76.089
Protein$^-$	1,874,703	2,204,068	85.057
University$^-$	74,638	79,213	94.224

Table 2 The information of caterpillars

Dataset	#	n	d	h	λ	β
N-glycans	514	6.40	1.84	4.22	2.18	4.50
all-glycans	7,984	4.74	1.49	3.02	1.72	2.84
CSLOGS	41,592	5.84	3.05	2.20	3.64	5.18
dblp$_{0.1}$	5,154	41.74	40.73	1.01	40.73	10.62
SwissProt	6,804	35.10	24.96	2.00	33.10	16.79
TPC-H°	8	8.63	7.63	1.00	7.63	8.63
Auction$^-$	259	4.23	3.00	0.71	3.57	4.29
Nasa$^-$	21,245	2.49	1.13	0.86	1.69	2.21
Protein$^-_{0.02}$	3,748	95.84	89.76	2.63	93.21	7.52
University$^-$	74,638	1.25	0.25	0.12	1.12	1.25

In experiments, we deal with caterpillars for N-glycans, all-glycans, CSLOGS, the largest 5,154 caterpillars (0.1%) in dblp (we refer to dblp$_{0.1}$), SwissProt, non-isomorphic caterpillars in TPC-H (we refer to TPC-H°), Auction$^-$, Nasa$^-$, the largest 3,748 caterpillars (0.02%) in Protein$^-$ (we refer to Protein$^-_{0.02}$) and University$^-$. Table 2 illustrates the information of such caterpillars. Here, # is the number of caterpillars, n is the average number of nodes, d is the average degree, h is the average height, λ is the average number of leaves and β is the average number of labels.

Table 3 The running time of computing distances τ_{CALN}, τ_{TAI} and τ_{ILST} (s)

Dataset	τ_{CALN}		τ_{TAI}		τ_{ILST}
	Total	Pair	Total	Pair	
N-glycans	114.51	8.69×10^{-4}	635.97	4.82×10^{-3}	1.10
all-glycans	13,406.00	4.21×10^{-4}	57,011.10	1.79×10^{-3}	176.85
$\mathrm{dblp}_{0.1}$	45,846.70	3.45×10^{-3}	6,363.79	4.79×10^{-4}	14,148.45
SwissProt	54,012.40	2.25×10^{-3}	76,345.60	3.30×10^{-3}	10,389.95
TPC-H°	8.95×10^{-3}	3.19×10^{-4}	5.51×10^{-3}	1.90×10^{-4}	9.00×10^{-3}
Auction^{-}	1.91	5.70×10^{-5}	4.03	1.21×10^{-4}	0.21

Table 4 The comparison of τ_{CALN} with τ_{ILST}

Dataset	#Pairs	$\tau_{\mathrm{CALN}} < \tau_{\mathrm{ILST}}$		$\tau_{\mathrm{CALN}} = \tau_{\mathrm{ILST}}$		$\tau_{\mathrm{CALN}} > \tau_{\mathrm{ILST}}$	
N-glycans	131,841	764	0.58%	131,077	99.42%	0	0%
all-glycans	31,868,136	40,002	0.13%	31,822,515	99.86%	5,599	0.02%
$\mathrm{dblp}_{0.1}$	13,279,281	0	0%	13,279,281	100%	0	0%
SwissProt	23,143,806	11,372	0.05%	23,132,434	99.95%	0	0%
TPC-H°	28	0	0%	28	100%	0	0%
Auction^{-}	33,411	0	0%	33,411	100%	0	0%

Next, we evaluate the running time to compute the caterpillar alignment distance τ_{CALN}, the edit distance τ_{TAI} and the isolated-subtree distance τ_{ILST}, Table 3 illustrates the running time to compute them for all the pairs of caterpillars in Table 2, where "total" denotes the total running time and "pair" denotes the running time for a pair of caterpillars. Since we cannot compute τ_{TAI} for CSLOGS, NASA^{-}, $\mathrm{Protein}^{-}_{0.02}$ and University^{-} within 1 day, we omit them.

Table 3 shows that, whereas the time complexity of computing τ_{CALN} is same as that of computing τ_{TAI} in theoretical (Theorem 2 and 4), the running time of computing τ_{CALN} for N-glycans, all-glycans, SwissProt and Auction^{-} (*resp.*, for $\mathrm{dblp}_{0.1}$ and TCP-H°) is smaller (*resp.*, larger) than that of computing τ_{TAI}. In particular, the running time of computing τ_{CALN} for N-glycans, all-glycans and Auction^{-} is much smaller than that of computing τ_{TAI}. The reason is that, for their data, the computation of τ_{CALN} has more cases to compare the set of leaves with an empty forest than the computation of τ_{TAI}, and then the sizes of trees or forests substituted in the reccurrences in τ_{CALN} is smaller than those in τ_{TAI}.

On the other hand, the running time of computing τ_{CALN} for $\mathrm{dblp}_{0.1}$ is much larger than that of computing τ_{TAI}. The reason is that the height of almost trees in these data is 1. Then, the running time depends on the number of branches in the reccurrences and the number of branches in τ_{CALN} is larger than that in τ_{TAI}.

Finally, we compare the caterpillar alignment distance τ_{CALN} with the isolated-subtree distance τ_{ILST}. Table 4 illustrates the number and the ratio of cases that $\tau_{\mathrm{CALN}} < \tau_{\mathrm{ILST}}$, $\tau_{\mathrm{CALN}} = \tau_{\mathrm{ILST}}$ and $\tau_{\mathrm{CALN}} > \tau_{\mathrm{ILST}}$ for all the pairs of caterpillars.

Table 5 The comparison of τ_{TAI} with τ_{CALN}

Dataset	#Pairs	$\tau_{\mathrm{TAI}} < \tau_{\mathrm{CALN}}$		$\tau_{\mathrm{TAI}} = \tau_{\mathrm{CALN}}$		$\tau_{\mathrm{TAI}} = \tau_{\mathrm{CALN}} = \tau_{\mathrm{ILST}}$	
N-glycans	131,841	459	0.35%	131,382	99.65%	130,618	99.07%
all-glycans	31,868,136	37,104	0.12%	31,831,032	99.88%	31,789,562	99.75%
$\mathrm{dblp}_{0.1}$	13,279,281	0	0%	13,279,281	100%	13,279,2810	100%
SwissProt	23,143,806	5,933,179	25.64%	17,210,627	74.36%	17,210,627	74.36%
TPC-H°	28	0	0%	28	100%	28	100%
Auction⁻	33,411	0	0%	33,411	100%	33,411	100%

Table 4 shows that, whereas almost cases satisfy that $\tau_{\mathrm{CALN}} \leq \tau_{\mathrm{ILST}}$, the ratio that $\tau_{\mathrm{CALN}} < \tau_{\mathrm{ILST}}$ is very small. In particular, for all-glycans, there exist cases that $\tau_{\mathrm{CALN}} > \tau_{\mathrm{ILST}}$.

Table 5 illustrates the number and the ratio of cases that $\tau_{\mathrm{TAI}} < \tau_{\mathrm{CALN}}$, $\tau_{\mathrm{TAI}} = \tau_{\mathrm{CALN}}$ and $\tau_{\mathrm{TAI}} = \tau_{\mathrm{CALN}} = \tau_{\mathrm{ILST}}$ for all the pairs of caterpillars.

Table 5 shows that almost cases except SwissProt satisfy that $\tau_{\mathrm{TAI}} = \tau_{\mathrm{CALN}}$. In particular, for $\mathrm{dblp}_{0.1}$, TCP-H° and Auction^{-}, it holds that $\tau_{\mathrm{TAI}} = \tau_{\mathrm{CALN}} = \tau_{\mathrm{ILST}}$ for all the pairs of caterpillars. The reason is that almost pairs of caterpillars have the height 1 and then the distance coincides with the sum of the cost of the correspondence between the roots and the bag distance between the set of leaves.

On the other hand, for SwissProt, the ratio that $\tau_{\mathrm{TAI}} < \tau_{\mathrm{CALN}}$ is much larger than others, because SwissProt has many caterpillars whose height is 2 and whose number of leaves is large. In particular, by Table 4 and 5, 74.36%, 25.59% and 0.05% of pairs in SwissProt satisfy that $\tau_{\mathrm{TAI}} = \tau_{\mathrm{CALN}} = \tau_{\mathrm{ILST}}$, $\tau_{\mathrm{TAI}} < \tau_{\mathrm{CALN}} = \tau_{\mathrm{ILST}}$ and $\tau_{\mathrm{TAI}} < \tau_{\mathrm{CALN}} < \tau_{\mathrm{ILST}}$, respectively.

7 Conclusion and Future Works

In this paper, we have formulated the caterpillar alignment distance τ_{CALN} and designed the algorithm to compute the alignment distance τ_{CALN} between caterpillars in $O(h^2\lambda^3)$ time under the general cost function and in $O(h^2\lambda)$ time under the unit cost function. Then, we have given the experimental results of computing τ_{CALN} comparing with τ_{ILST}.

As a result, whereas the time complexity of computing τ_{CALN} is same as that of computing τ_{TAI} in theoretical, there exist cases that the running time of computing τ_{CALN} is much smaller or much larger than that of computing τ_{TAI}. Also, almost cases satisfy that $\tau_{\mathrm{CALN}} \leq \tau_{\mathrm{ILST}}$, the ratio that $\tau_{\mathrm{CALN}} < \tau_{\mathrm{ILST}}$ is very small and there exist slightly cases that $\tau_{\mathrm{CALN}} > \tau_{\mathrm{ILST}}$. Furthermore, whereas almost cases satisfy that $\tau_{\mathrm{TAI}} = \tau_{\mathrm{CALN}}$, there exist data such that $\tau_{\mathrm{TAI}} < \tau_{\mathrm{CALN}} = \tau_{\mathrm{ILST}}$ is larger than other data.

Since the proof in Theorem 7 is naïve, it is possible to improve the time complexity, together with that of computing the edit distance between multisets under the general cost function (Lemma 4), which is a future work.

In this paper, we have just designed the algorithm to compute τ_{CALN}, not to compute τ_{ALN}. Then, as stated in the last of Sect. 5, it is an important future work to analyze whether or not the problem of computing τ_{ALN} for caterpillars is tractable and, if so, then to design the algorithm to compute τ_{ALN} for caterpillars.

References

1. Akutsu, T., Fukagawa, D., Halldórsson, M.M., Takasu, A., Tanaka, K.: Approximation and parameterized algorithms for common subtrees and edit distance between unordered trees. Theoret. Comput. Sci. **470**, 10–22 (2013). https://doi.org/10.1016/j.tcs.2012.11.017
2. Chawathe, S.S.: Comparing hierarchical data in external memory. In: Proceedings of VLDB'99, pp. 90–101 (1999). https://doi.org/10.5555/645925.671669
3. Deza, M.M., Deza, E.: Encyclopedia of Distances, 4th edn. Springer (2016). https://doi.org/10.1007/978-3-662-52844-0
4. Gallian, J.A.: A dynamic survey of graph labeling. Electorn. J. Combin. DS6 (2018)
5. Hirata, K., Yamamoto, Y., Kuboyama, I.: Improved MAX SNP-hard results for finding an edit distance between unordered trees. In: Proceedings of CPM'11, LNCS, vol. 6661, pp. 402–415 (2011). https://doi.org/10.1007/978-3-642-21458-5_34
6. Ishizaka, Y., Yoshino, T., Hirata, K.: Anchored alignment problem for rooted labeled trees. In: JSAI-isAI Post-Workshop Proceedings, LNAI, vol. 9067, pp. 296-309 (2015). https://doi.org/10.1007/978-3-662-48119-6_22
7. Jiang, T., Wang, L., Zhang, K.: Alignment of trees—An alternative to tree edit. Theoret. Comput. Sci. **143**, 137–148 (1995). https://doi.org/10.1016/0304-3975(95)80029-9
8. Kuboyama, T.: Matching and learning in trees. Ph.D thesis, University of Tokyo (2007)
9. Lu, C.L., Su, Z.-Y., Yang, C.Y.: A new measure of edit distance between labeled trees. In: Proceedings of COCOON'01, LNCS, vol. 2108, pp. 338–348 (2001). https://doi.org/10.1007/3-540-44679-6_37
10. Muraka, K., Yoshino, T., Hirata, K.: Computing edit distance between rooted labeled caterpillars. In: Proceedings of FedCSIS'18, pp. 245–252 (2018). https://doi.org/10.15439/2018F179
11. Muraka, K., Yoshino, T., Hirata, K.: Vertical and horizontal distance to approximate edit distance for rooted labeled caterpillars. In: Proceedings of ICPRAM'19, pp. 590–597 (2019). https://doi.org/10.5220/0007387205900597
12. Selkow, S.M.: The tree-to-tree editing problem. Inf. Process. Lett. **6**, 184–186 (1977). https://doi.org/10.1016/0020-0190(77)90064-3
13. Tai, K.-C.: The tree-to-tree correction problem. J. ACM **26**, 422–433 (1979). https://doi.org/10.1145/322139.322143
14. Wang, J.T.L., Zhang, K.: Finding similar consensus between trees: an algorithm and a distance hierarchy. Pattern Recog. **34**, 127–137 (2001). https://doi.org/10.1016/50031-3203(99)00199-5
15. Yamamoto, Y., Hirata, K., Kuboyama, T.: Tractable and intractable variations of unordered tree edit distance. Int. J. Found. Comput. Sci. **25**, 307–330 (2014). https://doi.org/10.1142/50129054114500154
16. Yoshino, T., Hirata, K.: Tai mapping hierarchy for rooted labeled trees through common subforest. Theory Comput. Sys. **60**, 759–783 (2017). https://doi.org/10.1007/s00224-016-9705-1
17. Zhang, K.: A constrained edit distance between unordered labeled trees. Algorithmica **15**, 205–222 (1996). https://doi.org/10.1007/BF01975866
18. Zhang, K., Jiang, T.: Some MAX SNP-hard results concerning unordered labeled trees. Inf. Process. Lett. **49**, 249–254 (1994). https://doi.org/10.1016/0020-0190(94)90062-0
19. Zhang, K., Wang, J., Shasha, D.: On the editing distance between undirected acyclic graphs. Int. J. Found. Comput. Sci. **7**, 45–58 (1996). https://doi.org/10.1142/S0129054196000051

ICrA Over Ordered Pairs Applied to ABC Optimization Results

Olympia Roeva and Dafina Zoteva

Abstract In this paper eight differently tuned Artificial Bee Colony (ABC) algorithms are considered. The influence of the *population size*, *the maximum number of iterations* and the probability for choosing a food source (p_i) on the algorithm performance is investigated. Parameter identification of an *E. coli* fed-batch fermentation process model is used as a case study. The ABC algorithms with the best performance are defined based on the obtained numerical results. The considered ABC algorithms are compared applying InterCriteria Analysis. An additional knowledge about the ABC algorithms correlations is obtained as a result.

Keywords Workforce planning · Artificial Bee Colony · Metaheuristics · InterCriteria analysis

1 Introduction

InterCriteria Analysis (ICrA) [1] has been developed to support the decision making process in a multicriteria environment where a number of objects are evaluated according to some predefined criteria. ICrA is based on two mathematical formalisms. The arrays of data with the estimates of the objects according to the criteria are processed using the apparatus of index matrices (IM) [2]. As for the uncertainties in the process, they are handled using the intuitionistic fuzzy sets (IFS) [3]. The aim is to explore the nature of the involved criteria. The relation between each pair of criteria is estimated in the form of intuitionistic fuzzy pair (IFP) [4].

O. Roeva (✉) · D. Zoteva
Institute of Biophysics and Biomedical Engineering,
Bulgarian Academy of Sciences, Sofia, Bulgaria
e-mail: olympia@biomed.bas.bg

D. Zoteva
e-mail: dafy.zoteva@gmail.com

S. Fidanova (ed.), *Recent Advances in Computational Optimization*,
Studies in Computational Intelligence 920,
https://doi.org/10.1007/978-3-030-58884-7_7

Numerous applications of ICrA have been found in different areas of science and practice. Some of the studies are related to problems in the fields of medicine [5, 6], economy [7], neural networks [8], ecology [9], sports [10, 11], etc.

ICrA has been successfully applied to establish certain dependencies in model parameter identification procedures. The influence of some of the genetic algorithm (GA) parameters on the procedure of identifying parameters of a fermentation process model is examined in a series of papers. The population number and the number of iterations are studied in [12], the influence of the generation gap is investigated in [13]. A research on the mutation rate influence is published in [14]. A comparison of different pure and hybrid metaheuristic techniques applied to a model parameter identification is discussed in [15]. The relations between parameters of GA used in identification procedures are estimated with ICrA in [16]. The influence of the number of objects on the results of ICrA in identification procedures of model parameters is studied in [17].

An efficient and perspective population based optimization algorithm - Artificial Bee Colony (ABC)—is chosen as an object of this study. There are three continuous optimization algorithms based on the intelligent behaviour of the honeybees. Yang's Virtual bee algorithm [18] has been developed to optimize two-dimensional numeric functions. Pham's Bees algorithm [19] employs several control parameters. The ABC algorithm [20], developed by Karaboga, optimizes multi-variable and multi-modal numerical functions.

ABC optimization algorithm can be used efficiently in many optimization problems. ABC clustering algorithms are introduced in [21, 22]. A new search strategy is employed in the ABC algorithm for a protein structure prediction problem in [23]. A discrete ABC algorithm is proposed to solve the lot-streaming flow shop scheduling problem in [24]. A hybrid ABC algorithm with Monarch Butterfly Optimization as a mutation operator to replace the employee phase of the ABC algorithm is introduced in [25] for global optimization problems. An enhanced artificial bee colony optimization is considered in [26]. A comprehensive survey of advances of ABC and its applications is presented in [27].

ABC algorithm is used in the present research to solve a model parameter identification problem. As a case study, *E. coli* MC4110 fed-batch cultivation process model is considered. Differently tuned ABC optimization algorithms are examined in terms of ICrA in search of possible relations between them. In addition, the influence of ABC parameters on the algorithm performance is investigated.

The paper is organized in the following way. Section 2 presents the background of the standard ICrA. The background of the ABC optimization algorithm is given in Sect. 3. The formulation of the problem is discussed in Sect. 4. The discussions on the numerical results are presented in Sect. 5. The InterCriteria Analysis of the results is presented in Sect. 6. The concluding remarks are in Sect. 7.

2 InterCriteria Analysis Background

The theoretical background of the standard ICrA is briefly discussed here.

Let O denotes the set of all objects $O_1, O_2, \ldots, O_n$ which are being evaluated according to a given criterion, and $C(O)$ be the set of the estimated values of the objects against the criterion C, i.e.,

$$O \stackrel{\text{def}}{=} \{O_1, O_2, O_3, \ldots, O_n\},$$

$$C(O) \stackrel{\text{def}}{=} \{C(O_1), C(O_2), C(O_3), \ldots, C(O_n)\}.$$

Let an IM whose index sets consist of the names of the criteria for the rows and the objects for the columns be given. The elements of this IM are the estimates of the objects according to each of the given criteria. As a result of ICrA applied to the input IM, a new IM is constructed with index sets consisting of the names of the criteria only. The elements of this resulting IM are IFPs, defined as the degrees of "agreement" and "disagreement" between each pair of criteria. An IFP is an ordered pair of real non-negative numbers $\langle a, b\rangle$, where $a + b \leq 1$.

Let $x_i = C(O_i)$. Then the following set can be defined:

$$C^*(O) \stackrel{\text{def}}{=} \{\langle x_i, x_j\rangle | i \neq j \;\&\; \langle x_i, x_j\rangle \in C(O) \times C(O)\}.$$

Let each ordered pair $\langle x, y\rangle \in C^*(O)$ for a fixed criterion C fulfill exactly one of the three relations R, $\overline{R}$ and $\tilde{R}$:

$$\langle x, y\rangle \in R \Leftrightarrow \langle y, x\rangle \in \overline{R} \tag{1}$$

$$\langle x, y\rangle \in \tilde{R} \Leftrightarrow \langle x, y\rangle \notin (R \cup \overline{R}) \tag{2}$$

$$R \cup \overline{R} \cup \tilde{R} = C^*(O). \tag{3}$$

Further, R, $\overline{R}$ and $\tilde{R}$ are considered to be $>$, $<$ and $=$, respectively.

In order to find the "agreement" between two criteria, the vector of all internal comparisons of each criterion should be constructed. Following Eqs. (1)–(3), if the relation between x and y is known, so is the relation between y and x. Therefore, for the construction of the vector of internal comparisons only a subset of $C(O) \times C(O)$ needs to be considered. More precisely, only the lexicographically ordered pairs $\langle x, y\rangle$ are of interest.

Let $C_{i,j} = \langle C(O_i), C(O_j)\rangle$. Then the vector of internal comparisons for a criterion C is:

$$V(C) = \left\{C_{1,2}, C_{1,3}, \ldots, C_{1,n}, C_{2,3}, C_{2,4}, \ldots, C_{2,n}, C_{3,4}, \ldots, C_{3,n}, \ldots, C_{n-1,n}\right\}.$$

The number of its elements is $n(n-1)/2$.

Let $V(C)$ is replaced by $\hat{V}(C)$, where its k-th component $(1 \le k \le n(n-1)/2)$ is defined as follows:

$$\hat{V}_k(C) = \begin{cases} 1, & \text{iff } V_k(C) \in R, \\ -1, & \text{iff } V_k(C) \in \overline{R}, \\ 0, & \text{otherwise.} \end{cases}$$

The degree of "agreement" $\mu_{C,C'}$ between two criteria C and C' is determined as the number of matching non-zero components of the respective vectors (divided by the length of the vector for normalization purposes). The degree of "disagreement" $\nu_{C,C'}$ is the number of components of opposing signs in the two vectors, again normalized.

The pair $\langle \mu_{C,C'}, \nu_{C,C'} \rangle$ is an IFP and $\mu_{C,C'} = \mu_{C',C}$ and $\nu_{C,C'} = \nu_{C',C}$. For most of the obtained pairs $\langle \mu_{C,C'}, \nu_{C,C'} \rangle$, the sum $\mu_{C,C'} + \nu_{C,C'} = 1$. However, there may be some pairs, for which this sum is less than 1. The difference

$$\pi = 1 - (\mu_{C,C'} + \nu_{C,C'})$$

is considered as a degree of "uncertainty".

There are three software implementations of the ICrA algorithms at present. The first implementation of the classic algorithm is in the Matlab environment. The script uses CSV files with arbitrary dimensions of the input matrix. Some of the earlier studies of the ICrA use this Matlab implementation [12, 28]. A stand-alone software application has been developed later by Mavrov [29, 30]. The application with a simple user friendly interface works only in Windows environment. The input data can be stored in .xls, .xlsx files or in tab-delimited .txt files.

A new cross-platform software implementing ICrA, called ICrAData, is proposed in [31]. ICrAData employs five different algorithms for calculating the intercriteria relations, according to the current development of the ICrA theory [32] and has no limitation on the number of objects. The present research utilizes this software implementation.

3 Artificial Bee Colony Optimization Algorithm

ABC is a swarm algorithm which simulates the foraging behaviour of the honey bees [20]. ABC is one of the competitive optimization algorithms, highly effective for multi-dimensional numerical problems.

The search involves three kinds of bees: employed bees, onlookers and scouts.

Each of the employed bees is associated with a specific food source. Each food source is considered a potential solution of the optimization problem. The evaluation or the fitness of a certain solution corresponds to the amount of the nectar in the food source. The population size is usually twice as large as the number of the employed bees or otherwise the number of the food sources associated with them. The other half of the population are onlookers.

The algorithm parameters: *population size* (NP), *number of the food sources* (SN), *number of parameters* (D, dimension of the search space), *maximum number of trials before abandoning a food source* (limit) and *maximum number of iterations or cycles* (MCN) are set in the initialization phase of the algorithm.

The initial population of food sources is generated then. Each food source is a D-dimensional vector $x_i = \{x_i^1, x_i^2, x_i^D\}$, limited by a lower $x_{\min}$ (Lb) and an upper $x_{\max}$ bounds (Ub), defined by the optimization problem. The count of the food sources is precisely SN. Equation (4) is used for generating each of the food sources:

$$x_i^j = x_{\min}^j + rand(0, 1)(x_{\max}^j - x_{\min}^j), \quad (4)$$

where $i \in [1; SN], j \in [1; D], x_{\min}^j$ and $x_{\max}^j$ are the lower and the upper bounds of the dimension j. The number of trials for each of the generated food sources is set to 0. This counter is incremented each time the food source associated with it is replaced by a better one.

Each iteration of the algorithm includes the following three phases: an employed bees phase, an onlookers phase and a scouts phase.

During the *employed bees phase* the employed bees are sent to their distributed food sources. The employed bees search for new food sources around the one stored in their memory, based on the Eq. (5):

$$v_{ij} = x_{ij} + \phi_{ij}(x_{ij} - x_{kj}), \quad (5)$$

where j is a random integer number in the range $[1; D]$, k is randomly selected index, $k \in [1; SN]$, $k \neq i$. $\phi_{ij} \in [-1; 1]$ is a random number. If the value generated by this operation exceeds its predefined boundaries, it should be pushed back into its limits. The quality of the new solution is evaluated then. A greedy selection is applied between x_i and v_i, the better solution is selected based on the fitness values. If v_i is better, it will replace x_i in the memory of the corresponding employed bee and the count of the trials will be set back to 0. Otherwise the old food source is kept in the memory, but the count of the trials is incremented by 1.

During the *onlookers phase* the onlookers choose a food source depending on the probability value p_i associated with that food source. It is calculated by [33]:

$$p_i = \frac{f_i}{\sum_{n=1}^{SN} f_n}, \quad (6)$$

where f_i is the fitness value of the solution i and it is proportional to the nectar amount of the food source in position i.

In order to increase the probability of the lower fitness individuals to be selected, an artificial onlooker bee chooses a food source depending on the probability [23]:

$$p_i = \frac{0.9 f_i}{\max(f_i)} + 0.1. \tag{7}$$

The onlookers search around the chosen food source, based on Eq. (5), evaluate the quality of the new source and apply greedy selection like the employed bees.

The *scouts phase* begins after the employed bees and the onlookers finish their search. The value of the trials for each food source is compared to the limit, set in the initialization phase. If the value of the trials exceeds the limit, the corresponding food source is abandoned. The abandoned food source is replaced by a food source generated randomly by Eq. (4), i.e., a food source found by a scout.

The algorithm stops when the predefined *maximum number of iterations* (MCN) is reached.

4 Case study: *E. coli* Fed-Batch Cultivation Process

The mathematical model of the *E. coli* fed-batch process is presented by the following non-linear differential equations system:

$$\frac{dX}{dt} = \mu_{\max} \frac{S}{k_S + S} X - \frac{F_{in}}{V} X \tag{8}$$

$$\frac{dS}{dt} = -\frac{1}{Y_{S/X}} \mu_{\max} \frac{S}{k_S + S} X + \frac{F_{in}}{V} (S_{in} - S) \tag{9}$$

$$\frac{dV}{dt} = F_{in} \tag{10}$$

where:

- X is biomass concentration, [g/l];
- S is substrate concentration, [g/l];
- F_{in} is feeding rate, [l/h];
- V is bioreactor volume, [l];
- S_{in} is substrate concentration in the feeding solution, [g/l];
- $\mu_{\max}$ is the maximum value of the specific growth rate, [1/h];
- k_S is saturation constant, [g/l];
- $Y_{S/X}$ is yield coefficient, [-].

For the parameter estimation problem, real experimental data of an *E. coli* MC4110 fed-batch cultivation process are used. Measurements of the biomass and glucose concentrations are used in the identification procedure. The parameter vector that should be identified is $p = [\mu_{\max}\ k_S\ Y_{S/X}]$.

The objective function is presented as a minimization of a distance measure J between experimental data and model predicted values of the process variables:

$$J = \sum_{i=1}^{m} \left(X_{\exp}(i) - X_{\mathrm{mod}}(i)\right)^2 + \sum_{i=1}^{n} \left(S_{\exp}(i) - S_{\mathrm{mod}}(i)\right)^2 \to \min$$

where

- m and n are the experimental data dimensions;
- $X_{\exp}$ and $S_{\exp}$ are the available experimental data for biomass and substrate;
- X_{mod} and S_{mod} are the model predictions for biomass and substrate for a given model parameter vector, $p = [\mu_{\max}\ k_S\ Y_{S/X}]$.

5 Numerical Results

Eight differently tuned ABC algorithms have been investigated. Different *population size* (NP), *maximum number of iterations* (MCN) and probability function p_i are applied to estimate *E. coli* model parameters. A series of 30 runs for each differently tuned ABC algorithm have been performed because of the stochastic characteristics of the applied algorithm. The used ABC algorithm parameters and functions are summarized in Table 1.

In order to compare the results of the identification of model parameters, the average, the best and the worst results of the 30 runs, for the model parameters ($\mu_{\max}$, k_S and $Y_{S/X}$), J value and the execution time (T) have been observed. The results obtained for the objective function and the computation time are summarized in Tables 2 and 3.

Table 1 ABC algorithm—Parameters and functions

ABC algorithms	Parameters		Probability, p_i
	NP	*MCN*	
ABC1	60	500	Eq. (6)
ABC2	40	400	
ABC3	40	200	
ABC4	20	400	
ABC5	60	500	Eq. (7)
ABC6	40	400	
ABC7	40	200	
ABC8	20	400	

Table 2 Results from ABC algorithm—Objective function values and standard deviation

ABC algorithms	Objective function value J			
	Average	Best	Worst	STD
ABC1	**4.4149**	**4.3001**	4.5076	0.0517
ABC2	**4.4568**	**4.3023**	4.5688	0.0671
ABC3	4.5094	4.3758	4.6248	0.0602
ABC4	4.5124	4.3785	4.6297	0.0577
ABC5	**4.4563**	4.3624	4.5329	0.0456
ABC6	4.4682	4.3774	4.5491	0.0523
ABC7	4.4907	**4.3268**	4.6032	0.0574
ABC8	4.4921	**4.3195**	4.5981	0.0643

Table 3 Results from ABC algorithm—Computation time

ABC algorithms	Computation time T		
	Average	Best	Worst
ABC1	534.4448	530.4688	538.4688
ABC2	286.3557	277.3438	291.8906
ABC3	146.7448	141.5313	149.5156
ABC4	146.7922	143.2344	150.6406
ABC5	544.8828	536.3594	560.9688
ABC6	280.7047	276.7813	288.6406
ABC7	146.1297	141.0000	149.9219
ABC8	149.2307	147.5156	152.1719

The results (both average and best), presented in Tables 2 and 3 show that the ABC1 algorithm finds the solution with the highest quality, e.g. the smallest objective function value ($J = 4.3001$). A similar result is shown by the ABC2 algorithm ($J = 4.3023$). These two ABC algorithms operate with large number of population and iterations.

Considering the best results, when the second form for calculating the probability (Eq. 2) is used, the worst results (ABC5 and ABC6) are observed in the case of large number of population and cycles. As it can be seen, the results of ABC7 and ABC8 are even better than the results of ABC3 and ABC4.

6 InterCriteria Analysis of the Results

In order to obtain some additional knowledge about the investigated here 8 ABC algorithms, ICrA is applied. The 8 IMs for all suggested ABC algorithms are constructed as follows:

$ABC1 =$

	run1	run2	...	run29	run30
μ_{max}	0.4779	0.4970	...	0.4887	0.4941
k_S	0.0096	0.0129	...	0.0120	0.0136
$Y_{S/X}$	2.0208	2.0224	...	2.0200	2.0205
J	4.4349	4.4062	...	4.4445	4.3587

$ABC2 =$

	run1	run2	...	run29	run30
$\mu_{\max}$	0.4699	0.4980	...	0.4733	0.4749
k_S	0.0083	0.0127	...	0.0085	0.0092
$Y_{S/X}$	2.0209	2.0211	...	2.0213	2.0205
J	4.5688	4.5560	...	4.3763	4.4304

etc.

The IMs are constructed using the obtained numerical values of the three model parameters ($\mu_{\max}$, k_S and $Y_{S/X}$) and the corresponding objective function value J.

In the first step, ICrA has been applied to each of the constructed ABC_s, $s = [1, 8]$ IM. The 30 runs are considered as objects and the three model parameters and objective function value J—as criteria. Thus, the relations between the model parameters and J for each ABC algorithm are obtained: $\mu_{max} - k_S$, $\mu_{\max} - Y_{S/X}$, $\mu_{\max} - J$, $k_S - Y_{S/X}$, $k_S - J$, $Y_{S/X} - J$.

ICrA has been applied again on these resulting 8 sets of 6 criteria pairs. In this case the 6 criteria pairs are the objects and the 8 ABC algorithms are criteria. Finally, correlations between the 8 investigated ABC algorithms have been obtained.

The ICrA results have been analyzed based on the proposed in [34] consonance and dissonance scale. The scheme for defining the consonance and dissonance between each pair of criteria is presented in Table 4.

All results, based on application of μ-biased ICrA [31], are visualized in Fig. 1 within the specific triangular geometrical interpretation of IFSs, thus allowing to order these results simultaneously according to the degrees of "agreement" $\mu_{C,C'}$ and "disagreement" $\nu_{C,C'}$ of the IFPs. Figure 1 shows that the resulting degrees of "uncertainty" $\pi_{C,C'}$ are very small for every criteria pairs.

Three groups of the obtained results are considered here: (1) ABC algorithms that are in positive consonance (Table 5), (2) ABC algorithms that are in weak positive consonance (Table 6) and (3) ABC algorithms that are in dissonance (Table 7).

ICrA shows that three of the ABC algorithms pairs have the strongest correlation among all the considered pairs. The pairs ABC1-ABC2, ABC1-ABC7 and ABC3-ABC5 have $\mu_{C_i,C_j} = 0.93$. As a confirmation of the previous discussion the best performed algorithms (ABC1 and ABC2) using Eq. (6) show higher correlation. As expected, ABC1 is correlated with ABC7, and ABC3 with ABC5. The good performance of ABC5 and ABC7 algorithms is also established by ICrA.

Some of the ABC algorithm pairs in weak positive consonance are pairs with the same probability function, such as ABC2-ABC3 or ABC3-ABC4 or ABC7-ABC8.

Table 4 Consonance and dissonance scale [34]

Interval of $\mu_{C,C'}$	Meaning
(0–0.05]	Strong negative consonance (SNC)
(0.05–0.15]	Negative consonance (NC)
(0.15–0.25]	Weak negative consonance (WNC)
(0.25–0.33]	Weak dissonance (WD)
(0.33–0.43]	Dissonance (D)
(0.43–0.57]	Strong dissonance (SD)
(0.57–0.67]	Dissonance (D)
(0.67–0.75]	Weak dissonance (WD)
(0.75–0.85]	Weak positive consonance (WPC)
(0.85–0.95]	Positive consonance (PC)
(0.95–1]	Strong positive consonance (SPC)

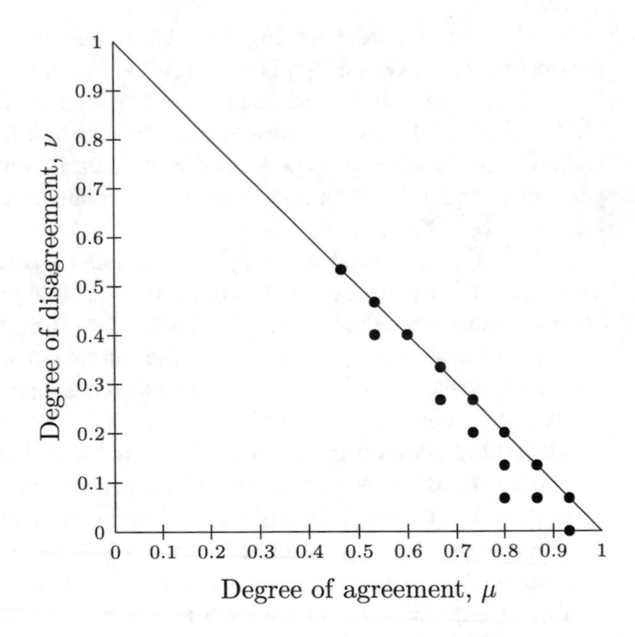

Fig. 1 Intuitionistic fuzzy triangle representation of the obtained results

Others use different probability functions, such as ABC4-ABC8 or ABC1-ABC5. Moreover, these ABC algorithms have the same values of NP and MCN.

The ABC algorithms that are in dissonance are considered in the last group. Most of them are algorithms that use different probability functions. There are ABC algorithms which use the same probability function but show different performance. For example, ABC6-ABC8 or ABC5-ABC6. Only the ABC algorithms which use probability Eq. (7) show such behavior. The results show that the ABC algorithms with different probability functions have different performance when varying the parameters NP and MCN. When the probability function Eq. (6) is used, the increase

Table 5 Results from ICrA—Positive consonance

ABC algorithms	Parameters		Probability, p_i	Criteria pair	μ_{C_i,C_j}
	NP	*MCN*			
ABC1	60	500	Eq. (6)	ABC1-ABC2	0.93
ABC2	40	400		ABC1-ABC7	0.93
ABC3	40	200		ABC3-ABC5	0.93
ABC4	20	400		ABC1-ABC4	0.87
ABC5	60	500	Eq. (7)	ABC2-ABC4	0.87
ABC6	40	400		ABC2-ABC7	0.87
ABC7	40	200		ABC3-ABC7	0.87
ABC8	20	400		ABC3-ABC8	0.87

Table 6 Results from ICrA—Weak positive consonance

ABC algorithms	Parameters		Probability, p_i	Criteria pair	μ_{C_i,C_j}
	NP	*MCN*		ABC1-ABC3	0.80
ABC1	60	500	Eq. (6)	ABC1-ABC5	0.80
ABC2	40	400		ABC2-ABC3	0.80
ABC3	40	200		ABC2-ABC5	0.80
ABC4	20	400		ABC3-ABC4	0.80
ABC5	60	500	Eq. (7)	ABC4-ABC5	0.80
ABC6	40	400		ABC4-ABC7	0.80
ABC7	40	200		ABC4-ABC8	0.80
ABC8	20	400		ABC5-ABC7	0.80
				ABC5-ABC8	0.80
				ABC7-ABC8	0.80

Table 7 Results from ICrA

ABC algorithms	Parameters		Probability, p_i	Criteria pair	μ_{C_i,C_j}
	NP	*MCN*		ABC1-ABC6	0.73
ABC1	60	500	Eq. (6)	ABC1-ABC8	0.73
ABC2	40	400		ABC2-ABC8	0.73
ABC3	40	200		ABC2-ABC6	0.67
ABC4	20	400		ABC6-ABC7	0.67
ABC5	60	500	Eq. (7)	ABC4-ABC6	0.60
ABC6	40	400		ABC3-ABC6	0.53
ABC7	40	200		ABC5-ABC6	0.53
ABC8	20	400		ABC6-ABC8	0.47

of NP and MCN values leads to a better algorithm performance. Whereas, in the case of probability function Eq. (7) no such correlation is observed. There are better results in both cases: of smaller values of the number of population and number of iterations—ABC7, and of bigger values—ABC1.

7 Conclusion

- Differently tuned ABC algorithms are investigated in order to define the influence of the *population size* (NP), *the maximum number of iterations* (MCN) and the probability for choosing a food source (p_i) on the algorithm performance.
- The 8 ABC algorithms are applied to parameter identification of an *E. coli* fed-batch fermentation process model.
- Based on the obtained results, the ABC algorithms with the best performance are defined.
- Further, by applying ICrA, more knowledge about the identification results of all 8 ABC algorithms is sought.
- The obtained results by ICrA confirm the previous choice of best performed ABC algorithms (ABC1 and ABC2) and give an additional knowledge about the relation and correlation between the investigated ABC algorithms.

Acknowledgements Work presented here is supported by the Bulgarian National Scientific Fund under Grant KP-06-N22/1 "Theoretical Research and Applications of InterCriteria Analysis".

References

1. Atanassov, K., Mavrov, D., Atanassova, V.: Issues Intuitionistic Fuzzy Sets Gener. Nets. Inter-Criteria decision making: a new approach for multicriteria decision making. Based on index matrices and intuitionistic fuzzy sets **11**, 1–8 (2014)
2. Atanassov K.: Index matrices: towards an augmented matrix calculus. Stud. Comput. Intell. **573** (2014)
3. Atanassov, K.: On Intuitionistic Fuzzy Sets Theory. Springer, Berlin (2020)
4. Atanassov, K., Szmidt, E., Kacprzyk, J.: On intuitionistic fuzzy pairs. Issues Intuitionistic Fuzzy Sets Gener. Nets **19**(3), 1–13 (2013)
5. Zaharieva, B., Doukovska, L., Ribagin, S., Michalikova, A., Radeva, I.: Intercriteria analysis of Behterev's kinesitherapy program. Notes Intuitionistic Fuzzy Sets **23**(3), 69–80 (2017)
6. Zaharieva, B., Doukovska, L., Ribagin, S., Radeva, I.: InterCriteria approach to Behterev's disease analysis. Notes Intuitionistic Fuzzy Sets **23**(2), 119–127 (2017)
7. Atanassova, V., Doukovska, L., Kacprzyk, J., Sotirova, E., Radeva, I., Vassilev, P.: Intercriteria analysis of the global competitiveness report: from efficiency to innovation-driven economies. J. Multiple-Valued Logic Soft Comput. 2018 (in press). ISSN 1542-3980
8. Sotirov, S., Sotirova, E., Melin, P., Castilo, O., Atanassov, K.: Modular neural network preprocessing procedure with intuitionistic fuzzy InterCriteria analysis method. In: Flexible Query Answering Systems, vol. 400. Advances in Intelligent Systems and Computing, 2015, pp. 175–186. Springer (2016)

9. Ilkova, T., Petrov, M.: Application of InterCriteria analysis to the Mesta River pollution modelling. Notes Intuitionistic Fuzzy Sets **21**(2), 118–125 (2015)
10. Antonov, A.: Dependencies between model indicators of general and special speed in 13–14 year old hockey players. Trakia J. (2020,in press)
11. Antonov, A.: Analysis and detection of the degrees and direction of correlations between Key indicators of physical fitness of 10–12-year-old hockey players. Int. J. Bioautom. **23**(3), 303–314 (2019). https://doi.org/10.7546/ijba.2019.23.3.000709
12. Pencheva, T., Angelova, M., Vassilev, P., Roeva, O.: InterCriteria analysis approach to parameter identification of a fermentation process model. Adv. Intell. Syst. Comput. **401**, 385–397 (2016)
13. Roeva, O., Vassilev, P.: InterCriteria analysis of generation gap influence on genetic algorithms performance. Adv. Intell. Syst. Comput. **401**, 301–313 (2016)
14. Roeva, O., Zoteva, D.: Knowledge discovery from data: InterCriteria analysis of mutation rate influence. Notes Intuitionistic Fuzzy Sets **24**(1), 120–130 (2018)
15. Fidanova, S., Roeva, O.: Comparison of Different Metaheuristic Algorithms based on InterCriteria Analysis, Journal of Computational and Applied Mathematics. Available online **7**, (2017). https://doi.org/10.1016/j.cam.2017.07.028
16. Pencheva, T., Angelova, M., Atanassova, V., Roeva, O.: InterCriteria analysis of genetic algorithm parameters in parameter identification. Notes Intuitionistic Fuzzy Sets **21**(2), 99–110 (2015)
17. Zoteva, D., Roeva, O.: InterCriteria analysis results based on different number of objects. Notes on Intuitionistic Fuzzy Sets **24**(1), 110–119 (2018)
18. Yang, X.S.: Engineering optimizations via nature-inspired virtual bee algorithms. In: Artificial Intelligence and Knowledge Engineering Applications: A Bioinspired Approach. Lecture Notes in Computer Science, vol. 3562, pp. 317–323 (2005)
19. Pham, D.T., Ghanbarzadeh, A., Koc, E., Otri, S., Rahim, S., Zaidi, M.: The bees algorithm. Technical Report. Manufacturing Engineering Centre, Cardiff University, UK (2005)
20. Karaboga, D.: An idea based on honeybee swarm for numerical optimization. Technical Report TR06, Engineering Faculty, Computer Engineering Department, Erciyes University (2005)
21. Karaboga, D., Ozturk, C.: A novel clustering approach: artificial bee colony (ABC) algorithm. Appl. Soft Comput. **11**(1), 652–657 (2011)
22. Zhang, C., Ouyang, D., Ning, J.: An artificial bee colony approach for clustering. Expert Syst. Appl. **37**(7), 4761–4767 (2010)
23. Li, Y., Zhou, C., Zheng, X.: The application of artificial bee colony algorithm in protein structure prediction. In: Pan, L., Pun, G., Prez-Jimnez, M.J., Song, T. (eds.) Bio-Inspired Computing Theories and Applications. Communications in Computer and Information Science, vol. 472, pp. 255-258. Springer, Berlin, Heidelberg (2014)
24. Pan, Q.K., Tasgetiren, M.F., Suganthan, P.N., Chua, T.J.: A discrete artificial bee colony algorithm for the lot-streaming flow shop scheduling problem. Inf. Sci. **181**(12), 2455–2468 (2011)
25. Ghanem, W.: Hybridizing artificial bee colony with monarch butterfly optimization for numerical optimization problems. In: First EAI International Conference on Computer Science and Engineering, 16 November 1112, 2016, Penang, Malaysia, http://eudl.eu/doi/10.4108/eai.27-2-2017.152257 (2016)
26. Tsai, P.-W., Pan, J.-S., Liao, B.-Y., Chu, S.-C.: Enhanced artificial bee colony optimization. Int. J. Innov. **5**(12B), 5081–5092 (2009)
27. Karaboga, D., Gorkemli, B., Ozturk, C., Karaboga, N.: A comprehensive survey: artificial bee col-ony (ABC) algorithm and applications. Artif. Intell. Rev. **42**(1), 2157 (2014)
28. Roeva, O., Vassilev, P., Fidanova, S., Paprzycki, M.: InterCriteria analysis of genetic algorithms performance. In: Fidanova, S. (eds.) Recent Advances in Computational Optimization. Studies in Computational Intelligence, vol. 655, pp. 235–260 (2016)
29. Mavrov, D.: Software for InterCriteria analysis: implementation of the main algorithm. Notes Intuitionistic Fuzzy Sets **21**(2), 77–86 (2015)
30. Mavrov, D., Radeva, I., Atanassov, K., Doukovska, L., Kalaykov, I.: InterCriteria software design: graphic interpretation within the intuitionistic fuzzy triangle. In: Proceedings of the Fifth International Symposium on Business Modeling and Software Design, pp. 279–283 (2015)

31. Ikonomov, N., Vassilev, P., Roeva, O.: ICrAData software for InterCriteria analysis. Int. J. Bioautom. **22**(1), 1–10 (2018)
32. Roeva, O., Vassilev, P., Angelova, M., Su, J., Pencheva, T.: Comparison of different algorithms for InterCriteria relations calculation. In: IEEE 8th International Conference on Intelligent Systems, pp. 567–572 (2016). ISBN 978-1-5090-1353-1
33. Karaboga, D., Akay, B.: A comparative study of artificial bee colony algorithm. Appl. Math. Comp. **214**, 108–132 (2009)
34. Atanassov, K., Atanassova, V., Gluhchev, G.: InterCriteria analysis: ideas and problems. Notes Intuitionistic Fuzzy Sets **21**(2), 81–88 (2015)

A Game Theoretical Approach for VLSI Physical Design Placement

Michael Rapoport and Tami Tamir

Abstract The physical design placement problem is one of the hardest and most important problems in micro chips production. The placement defines how to place the electrical components on the chip. We consider the problem as a combinatorial optimization problem, whose instance is defined by a set of 2-dimensional rectangles, with various sizes and wire connectivity requirements. We focus on minimizing the placement area and the total wire-length. We propose a local-search method for coping with the problem, based on natural dynamics common in game theory. Specifically, we suggest to perform variants of *Best-Response Dynamics* (BRD). In our method, we assume that every component is controlled by a selfish agent, who aim at minimizing his individual cost, which depends on his own location and the wire-length of his connections. We suggest several BRD methods, based on selfish migrations of a single or a cooperative of components. We performed a comprehensive experimental study on various test-benches, and compared our results with commonly known algorithms, in particular, with simulated annealing. The results show that selfish local-search, especially when applied with cooperatives of components, may be beneficial for the placement problem.

Keywords VLSI placement · Best-response dynamics · selfish local-search

1 Introduction

Physical design is a field in Electrical Engineering which deals with *very large scale integration* (VLSI). Specifically, physical design is the main step in the creation of *Integrated Circuit* (IC). The basic question is *how to place the electrical components on the chip*. This fundamental question became relevant with the invention of ICs in

This research was supported by The Israel Science Foundations (Grant No. 1036/17).

M. Rapoport · T. Tamir (✉)
School of Computer Science, The Interdisciplinary Center, Herzliya, Israel
e-mail: tami@idc.ac.il

S. Fidanova (ed.), *Recent Advances in Computational Optimization*, Studies in Computational Intelligence 920,
https://doi.org/10.1007/978-3-030-58884-7_8

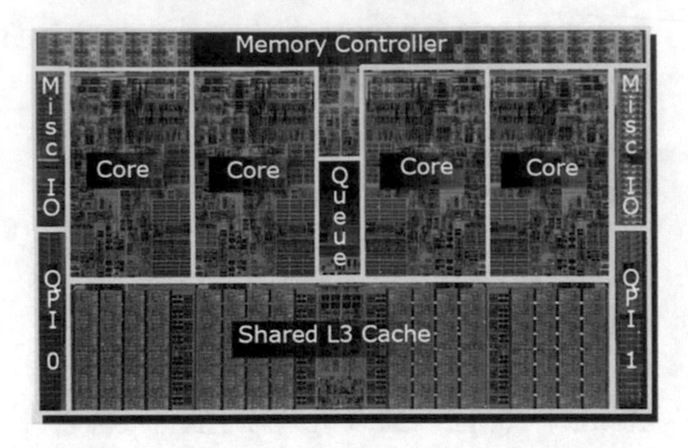

Fig. 1 Intel i7 processor placement

1958 [19], and remains critical our days with the development of micro-electricity. Recent developments in micro-electricity enables transistors to reach the size of nanometers, thus a single chip can accommodate thousands of components of different sizes and dispersed connectivity. Bad layout of electrical components leads to expensive production and poor performance. Figure 1 presents the Intel i7 processor [10], and demonstrates how efficient design is crucial in enabling the accommodation of many components on a small area.

The complexity of *VLSI physical design* led to the establishment of a design process, in which the problem is divided into several steps, each is an independent NP-complete problem. The most fundamental steps are: (i) *Floorplan:* choose the area of the chip and decide the positions of the building blocks of the chip, (i.e., in Intel processor: cores, graphic processor, cache, memory controller). (ii) *Placement:* Each of the above mentioned building blocks consists of several components. These components should be placed in a way that minimizes area and wire-length. (iii) *Signal and Clock Routing:* route the wires via components white space, which is an extra area assigned for wiring.

In this work, we focus on the *Placement* problem. The floorplan is usually performed manually, and the signal and clock routing is more of a production engineering problem which is tackled using different tools.

Several common methods for coping with the Placement problem are based on local-search. We propose a new class of local-search algorithms that consider the problem as a game played among the components, where each component corresponds to a selfish player who tries to maximize his own welfare. Our algorithms are different from other algorithms based on local search in the way they explore the solution space. Every solution is associated with a global cost, and every component is associated with its individual cost, which is based on its own placement and connections. We move from one solution to another if this move is selfishly beneficial for a single component or for a small cooperative of components, without considering the effect on the global cost.

In this paper we first review the placement problem, and survey some of the existing techniques to tackle it, in particular local-search algorithms and Simulated

Annealing. We then describe our new method of performing selfish Best-Response Dynamics (BRD). In order to evaluate this method, we performed an extensive experimental study in which we simulated and tested several variants of BRD on various test-benches. Our results show that BRD may produce a quick and high quality solution.

1.1 The Placement Problem

The Placement process determines the location of the various circuit components within the chip's core area. The problem, and even simple subclasses of it, were shown to be NP-complete by reductions to the *bin packing* and the *rectangle packing* problems [14, 15]. Moreover, a reduction to the *Quadratic assignment problem* shows that achieving even a constant approximation is NP-hard [9].

Bad placement not only reduces the chip's performance, but might also make it non-manufacturable by forcing very high wire-length, lack of space, or failing timing/power constraints. As demonstrated in Fig. 2, good placement involves an optimization of several objectives that together ensure the circuit meets its performance demand [4, 17]:

1. *Area*: Minimizing the total area used to accommodate the components reduces the cost of the chip and is crucial for the production.
2. *Total wire-length*: Minimizing the total length of the wires connecting the components is the primary objective of the physical design. Long wires require the insertion of additional buffering, to insure synchronization between the components. Short wires decrease the power consumption and the system's leakage.
3. *Wire intersection*: Our days, wire intersection is allowed as long as a single wire does not have more than a predefined number of intersections. The manufactur-

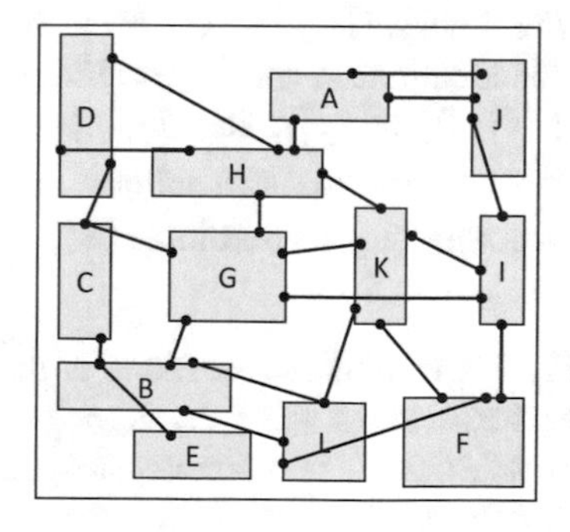

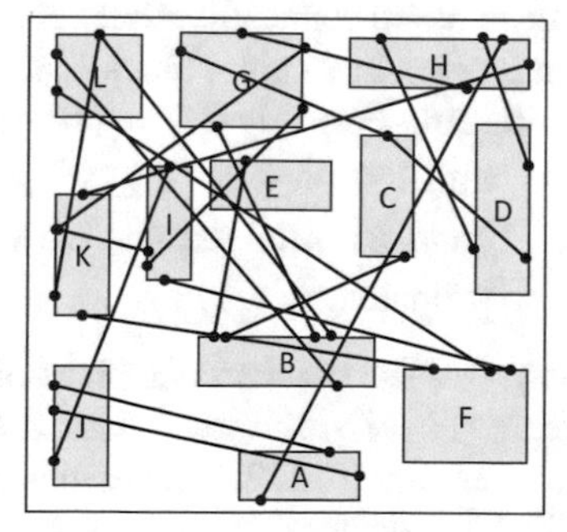

Fig. 2 An example of a good placement (left) versus a bad placement (right). In the good placement the wires are shorter and there is almost no congestion. In this example, the area of both placements is the same

ing process enables several routing layers. Nevertheless, a good layout avoids unnecessary intersections.
4. *Timing*: The timing cycle of a chip (clock frequency) is determined by the delay induced by the longest wire, usually referred to as the critical path.

Our work considers the initial placement calculation, denoted *global placement*. This stage is followed by the *detailed placement* stage, in which the global placement results are put into use and the cells are actually placed on the die. The detailed placement stage includes small changes to solve local issues such as wire congestion spots, remaining overlaps, layout constrains (such as via locations), connecting to the die pinout, etc.

In the global placement stage, several parameters are optimized. We focus on the total wire-length and placement area. By adjusting the cost function associated with each configuration, our method enables considering additional parameters such as wire congestion, critical path length, and more.

1.2 Formal Description of the Placement Problem

We describe the placement problem as a combinatorial optimization problem. The components composing the problem are represented by 2-dimensional rectangles denoted *blocks*. In the placement, they can be rotated by 90°, 180° or 270°, but not mirrored. The sides of the assigned blocks must be parallel to each other, and to the bounding area. Whenever we refer to a location in a block, we let (0, 0) be the bottom-left corner, and every other point in the block is given by its (x, y) coordinates with respect to this corner.

Formally, an instance of the problem is defined by:

1. A set of n blocks $\{B_1, B_2, \dots, B_n\}$ to be placed on the chip. Every block $1 \leq i \leq n$, has associated width w_i and height h_i.
2. A list of required connections between the blocks, $\{N_1, N_2, \dots, N_m\}$. Every connection is given by a pair of blocks, and the locations in which these blocks should be connected. Formally $N_j = \langle B_j^1, x_j^1, y_j^1, B_j^2, x_j^2, y_j^2 \rangle$, for $0 \leq x_j^1 \leq w_j^1, 0 \leq y_j^1 \leq h_j^1$ and $0 \leq x_j^2 \leq w_j^2, 0 \leq y_j^2 \leq h_j^2$, corresponds to a request to connect blocks B_j^1 and B_j^2, such that the wire is connected to coordinate (x_j^1, y_j^1) in B_j^1 and to coordinate (x_j^2, y_j^2) in B_j^2.

The output of the problem is a placement F given by the locations of the blocks on the plane $\{L_1, L_2, \dots, L_n\}$, such that for every $1 \leq i \leq n$, $L_i = (x_i, y_i, r_i)$. The parameter $r_i \in \{0, 1, 2, 3\}$ specifies how block B_i is rotated corresponding to $\{0, 90, 180, 270\}$ degrees. Note that a rotation by 180° is not equivalent to not rotating at all, since the location of the required connections is also rotated. Formally, block B_i is placed in the rectangle whose diagonal endpoints are (x_i, y_i) (this corner is independent of the value of r_i), and $(x_i + w_i, y_i + h_i)$ if $r_i = 0$, or $(x_i + h_i, y_i + w_i)$ if $r_i = 1$, or $(x_i - w_i, y_i - h_i)$ if $r_i = 2$, or $(x_i - h_i, y_i + w_i)$ if $r_3 = 1$.

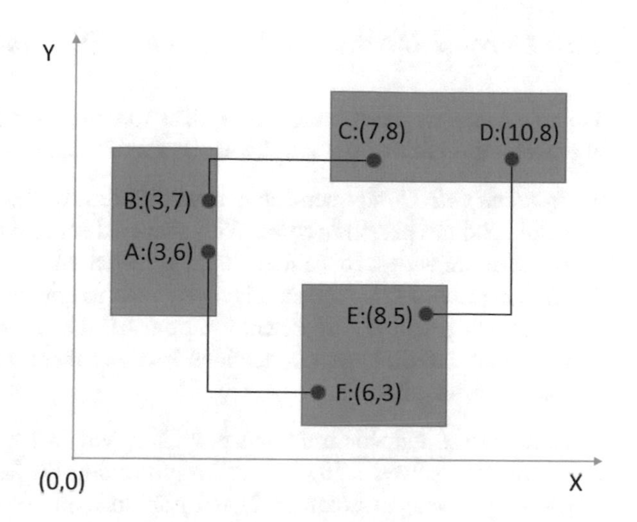

Fig. 3 An example of wire-length calculation. There are three connections between the pairs of points $\{A, F\}, \{D, E\}$ and $\{B, C\}$. The total wire-length is $(|X_A - X_F| + |Y_A - Y_F|) + (|X_E - X_D| + |Y_E - Y_D|) + (|X_B - X_C| + |Y_B - Y_C|) = (3 + 3) + (2 + 3) + (4 + 1) = 16$

A placement is legal if no two blocks overlap, that is, the rectangles induced by L_{i_1} and L_{i_2} are disjoint for all $i_1 \neq i_2$.

This condition may be relaxed a bit in the global placement stage, and allow small percentage of overlaps area. These overlaps are resolved later during the detailed placement stage.

The *bounding box* of a Placement F, is the minimum axis-aligned rectangle which contains all the blocks. The *area* of a placement F is the area of the bounding box, and is denoted $A(F)$.

The blocks' location, together with the required connections, induce the wire-length of a placement. Formally, assume that blocks B_1 and B_2 are located in L_1 and L_2, respectively, and let $N_j = \langle B_j^1, x_j^1, y_j^1, B_j^2, x_j^2, y_j^2 \rangle$. We first calculate the actual coordinates of the connection points, based on L_1, L_2, and the values of $\langle x_j^1, y_j^1 \rangle$ and $\langle x_j^2, y_j^2 \rangle$. Let $\langle \hat{x_j^1}, \hat{y_j^1} \rangle$ and $\langle \hat{x_j^2}, \hat{y_j^2} \rangle$ be the points we need to connect. The wire-length associated with N_j, denoted $Len(N_j)$, is calculated in a way that fits the actual production process, in which all wires are parallel to the blocks and to the bounding area, that is, $Len(N_j) = \Delta X + \Delta Y = |\hat{x_j^1} - \hat{x_j^2}| + |\hat{y_j^1} - \hat{y_j^2}|$. The total wire-length of a Placement F is denoted $L(F)$, and is given by $L(F) = \sum_{j=1}^{m} Len(N_j(F))$. An example of wire-length calculation is given in Fig. 3.

The goal in the placement problem is to minimize $\alpha L(F) + (1 - \alpha)A(F)$ where the parameter $0 \leq \alpha \leq 1$ weight the importance of the two objectives. These days (as the number of components per chip rises) it is a very common practice to focus on the wire-length of the placement and only when finished optimizing the wire-length, perform small changes in order to gain better area result, with a minimal harm of the achieved wire-length. Thus, in our experiments (to be described in Sect. 3), we give a substantially higher weight to the wire-length.

1.3 Current Techniques for Efficient Placement

We now overview the common disciplines to handle the Placement problem. Some algorithms are tailored for simplified classes of instances. Specifically,

1. *Standard cell*: Components may have different width, but they all have the same height and are placed in rows. With over-cell routing the goal is to minimize the width of the widest row and the total wire-length.
2. *Gate array / FPGA*: The area is discretized to equally sized squares where each square is a possible component location. All the components have the same size and shape but different connections between them, the goal is to minimize the total wire-length.

Both classes induce simplified problems, which are still NP-hard, but can be approximately solved using Linear Programming [7, 22], Greedy Algorithms [2, 25], Slicing Tree representation [3], or by Divide and Conquer algorithms that allows temporal block overlaps [2, 7].

A different solution approach is to develop heuristics, usually with strong randomness involved. Most heuristics have no assumptions on the problem thus able coping with general instances. Heuristics have no performance guarantee but perform well in practice. Some heuristics were tailored for *Standard cell* and *Gate array* instances [8, 16, 24]. The most commonly used algorithm concept for placement is *simulated annealing* (SA) [20, 23]. Modern algorithms of our days are always compared against it and many of them are based on its concept. While SA is unlikely to find an optimal solution, it can often find a very good one. The name simulated annealing come from annealing in metallurgy, a technique involving heating and controlled cooling of a material to increase the size of its crystals and reduce their defects. Both are attributes of the material that depend on its thermodynamic free energy. Heating and cooling the material affects both the temperature and the thermodynamic free energy. The simulation of annealing as an approach for minimization of a function of large number of variables was first formulated in [13]. Many modern algorithms are based on the concepts of simulated annealing.

Additional widely used placement methods include (i) Force Directed Placement, in which the problem is transformed into a classical mechanics problem of a system of objects attached to springs [21], (ii) Placement by Partitioning, in which the circuit is recursively partitioned into smaller groups [2, 7], (iii) Numerical Optimization Techniques, based on equation solving and eigenvalue calculations [18, 25], and (iv) Placement by Genetic Algorithm, that emulates the natural process of evolution as a means of progressing toward optimum, [25]. Some of these methods are only suited for *Standard cell* or *Gate array* instances, and some are general. A survey of the above and of additional algorithms for placement can be found in [1, 11, 24].

2 Our Local-Search Method for Solving the Placement Problem

The main challenges involved in solving the Placement problem are the need to optimize several objectives simultaneously, and to achieve even a good approximate solution in reasonable time. Optimizing even a single objective is an NP-hard problem. Naturally, combining several objectives, that may be conflicting, makes the problem more challenging.

Our proposed method not only performs a good placement in a relatively short time, but also copes with the multiple objective challenge.

2.1 The Placement Problem as a Game

We propose to tackle the problem by a local-search algorithm, using natural dynamics common in *game theory*. Specifically, we suggest to perform variants of *Best-Response Dynamics* (BRD), assuming the components correspond to strategic selfish agents who strive to optimize their own welfare. In a BRD process, every agent (player) in turn, selects his best strategy given the strategies of the other players. In our game, the strategy space of a player consists of all the locations his component can be placed in, given the location of the other components. Players keep changing strategies until a Nash equilibrium of the game is reached. A Nash equilibrium is a strategy profile in which no player can benefit from changing his strategy [12]. A lot of attention is given to best-response dynamics in the analysis of non-cooperative games, as this is the natural method by which players proceed toward a NE. The common research questions are whether BRD converges to a NE, the convergence time, and the quality of the solution (e.g. [5, 6]).

BRD can also be performed with coordinated deviations. That is, in each step, a group of players, denoted a *cooperative*, moves simultaneously, such that their total cost is reduced. Note that in a coordinated deviation of a *cooperative*, unlike a *coalition*, some members of the cooperative may be hurt. The deviation is beneficial if the total members' cost is reduced.

In order to consider the placement problem as a game played by selfish agents, we need to associate a value, or cost, for each player in each possible configuration of the game. In our setting, players correspond to blocks and configurations correspond to placements. The BRD process is define with respect to a cost function that depends on the wire-length connected to the player's block, and the total placement area. The individual cost function is calculated for each block or cooperative, and is relevant only to the currently playing block or cooperative.

Recall that for a configuration F, the global cost of F is

$$Global\_cost(F) = \alpha L(F) + (1 - \alpha)A(F),$$

where $L(F)$ is the total wire-length, $A(F)$ is the bounding box area, and the parameter α is used to weight these two components of the cost function. In our algorithms, the global cost function, is used only to evaluate the final configuration—in order to compare different methods and to analyze the progress of the algorithms.

The *individual cost function* is used to evaluate the possible deviations of the currently playing block. For a single block B_i, let $L_{B_i}(F)$ denote the total wire-length of B_i''s connections. By definition, $L(F) = \frac{1}{2}\sum_{1\leq i\leq n} L_{B_i}(F)$. The total individual wire-length is divided by 2, since every wire is counted in the individual wire-length of its two endpoints. For a configuration F and a block B_i, the individual cost of B_i in F is defined as follows:

$$Ind\_cost(B_i, F) = \alpha \cdot (L_{B_i}(F))^2 + (1-\alpha) \cdot A(F).$$

Note that in the individual cost function, the corresponding wire-length is squared – for normalization with the area component.

Let Γ be a subset of the blocks. In order to evaluate configurations that are a result of a coordinated deviation, we define, for a cooperative Γ in a configuration F, the individual cost of Γ in F:

$$Ind\_cost(\Gamma, F) = \alpha \cdot \sum_{B_i\in\Gamma} (L_{B_i}(F))^2 + (1-\alpha) \cdot A(F).$$

Since finding the best response is NP-hard in most scenarios and particularly for coordinated deviation, we perform a better-response move, in which the player (or a cooperative) benefits, but not necessarily in the optimal way. In practice, we perform the best response move in a restricted search space. Also, in some algorithms, when there is no local improving step, we may perform a move which harms the cost function. Such moves result in a temporary worse state and are used in order to allow the algorithm to escape from local minima.

In our experiments, we compared our results with those achieved by *Simulated Annealing* (SA), *Greedy Local Search* (GLS) algorithm, based on hill climbing, and *Fast Local Search* (FLS) algorithm (faster version of the greedy local search). For each test-bench we run our algorithms as well as these algorithms, and compared the results. In this paper we only provide the comparison with SA, as it outperformed the other two local-search methods.

2.2 Search for a Solution over the Solution Space

Before presenting our algorithms we give an overview of the local search technique, and explain how the search for the solution is performed. A local-search algorithm performs a search over the solution space. Every possible solution (placement F) is associated with a score ($Global\_cost(F)$). The global cost function defines a

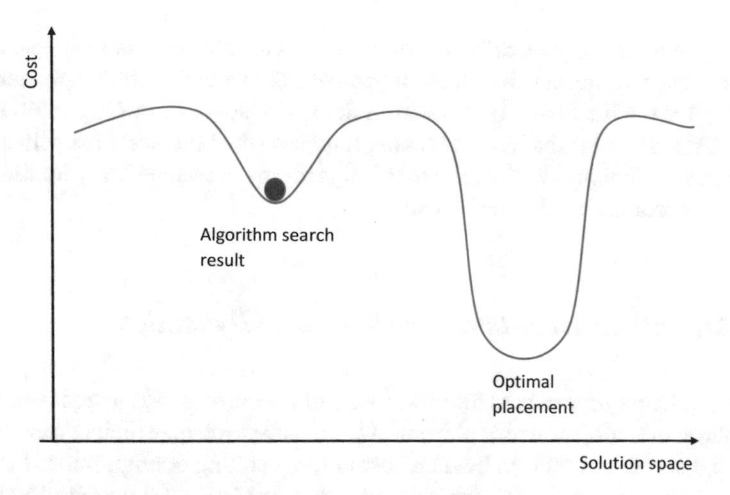

Fig. 4 A general description of a local-search method. The algorithm traverses the cost-placement multidimensional complex. Local search may end-up in a local minimum, unable to escape, thus also unable to find an optimal placement

placement-cost multidimensional complex, on which the algorithm advances. Each point on the complex is a placement and the complex includes all possible placements.

Every local-search algorithm moves on the placement-cost complex searching for a point corresponding to a placement having minimum cost. The local-search paradigm implies that the movement along the complex is almost continues. When the algorithm encounters a heap on the complex which it cannot pass, it may try to bypass it in order to continue the search in that direction.

The main challenge when applying such algorithms, is how to pick the next point to explore and how to decide when to stop the search. As demonstrated in Fig. 4, we can continue to search up to some point of worse cost but we do not know what awaits us further down the path of the search. We may attempt to remember each minimum we visit during the search and traverse different search paths from each local minimum detected. However, such methods perform a brute force search, which in turn results in exponential running time. Finding the global minimum means we have found an optimal solution for an NP-hard problem. Hence, such algorithms must have exponential running time (regardless of the algorithm's logic) unless P = NP.

The main difference between our algorithms and previous algorithms based on local search (in particular SA, GLS, FLS), is the way we evaluate each solution in the solution space, and the way we advance to the next solution in the search process. Previous algorithms calculate the cost function for the entire placement, while our algorithms base their progress on the individual cost of a block (or of a cooperative of blocks). The main goal of this work is to examine the quality of local-search algorithms for the placement problem, in which the search on the solution space is determined by only a single block (or a small cooperative), in a selfish manner, according to its individual cost-over-time curve. Our method does not use a global cost function; we present the global objective parameters (wire-length, total area, etc.) only for the analysis. As detailed in Sect. 2.3, some of our algorithms accept moves that harm a bit the individual cost function (thus breaking the selfishness to some extent). This possibility allows the algorithms to escape local minimum and the search becomes much more versatile.

2.3 Algorithms Based on Best-Response Dynamics

The BRD process proceeds as follows: Every block corresponds to a player. In every step a player or a cooperative of players have a chance to change their current location, in a way that reduces their individual cost in the resulting configuration. In some of our algorithms, a step increasing the individual cost may be accepted with some probability.

Every block can perform one or more of the following moves: {Up, Down, Left, Right, Rotate 90°, Rotate 180°, Rotate 270°}. A step is legal as long as the resulting configuration is legal.

We use three search methods for the block migrations:

1. *Best Response (BRD)*: Each block is controlled by a selfish player. Each player can perform one move per turn, the move is the best local move the block can perform to reduce its individual cost. The algorithm advances in rounds, where in each round, every player gets an opportunity to migrate. Players are allowed to perform only legal moves (no block-overlaps are created).
2. *Constant Perturbations (BRD-ConstPerb)*: In This variant of BRD, when a player does not have a legal improving move to perform, he may, with some non-negligible probability, choose a step which harms its individual cost. In our experiments we found 0.3 to be a good probability for accepting a worse state. It is small enough not to harm the selfishness on one hand, and allows the placement to escape local minima on the other.
3. *Relaxed Search (BRD-RlxSrch)*: This algorithm is another variant of BRD. The difference is that players can select illegal locations - that involve block overlaps. While blocks are not allowed to overlap beyond a reasonable limit in the final configuration, temporal overlaps may be fruitful. Our relaxed search allows overlaps with varying fines on the area of the overlap. The overlaps fine are added to the block's individual cost. The fines are increased every round - to encourage convergence to a final placement with hardly any overlaps. The Global placement stage can tolerate small overlaps, so the output is accepted if the final placement does include some overlaps.

Each of the above algorithms is ran in two variations: without and with *swap moves*. A swap move is a move in which the active block swaps places with some other block if the swap is legal and reduces the active block's individual cost, as well as the global cost function (this ensures we avoid recurrent swaps between a pair of blocks). Swap moves break the locality of the search and allows another method with which to escape local minima. Instead of attempting to escape a local minimum by accepting a worse state, the algorithm can escape a local minimum by jumping to a better, yet not local neighbor, state. Swap moves do not break the selfishness of our algorithms but rather only the locality, and only to some extent. As our experiments reveal, enabling swap moves improves the quality of the solution.

2.4 Coordinated Deviation of a Cooperative

Unlike a unilateral deviation, a coordinated deviation is initiated by a group of players, denoted a *cooperative*, who migrate simultaneously. Such a migration may harm the individual cost of some cooperative members (for example, if they give up good spots for other members), however, the total cost of the cooperative members is strictly reduced. When applying a coordinated deviation, we first determine the cooperative size and then the blocks composing it. A coordinated deviation is therefore defined by (i) the search method, (ii) the cooperative size method, and (iii) the cooperative member selection method.

We simulated three different methods for determining the cooperative's size:

1. *Increasing*: Starting from $k = 1$, after converging to a k-NE profile, which is stable against deviations of cooperatives of size k, we increase the active cooperative size to $k + 1$. We keep increasing the cooperative size up to a predefined limit.
2. *Iterating*: Each round has a different cooperative size, the sizes are incremented after each round, when the size reaches a predefined limit we reset the size to a single block.
3. *Random*: Each cooperative has a random size, the size is uniformly distributed between a single block and a predefined limit.

The cooperative's members are selected in the following way: we iterate over all the blocks, selecting a different *head block* of the cooperative in each round. The head block constructs a cooperative according to one of the following methods:

1. *Closest Connected blocks*: in every iteration we add to the cooperative a block with the shortest wire-length to some other block already in the cooperative.
2. *Farthest Connected blocks*: in every iteration we add to the cooperative a block with the longest wire-length to some other block already in the cooperative.
3. *Closest Geometrically blocks*: in every iteration we add to the cooperative a block with the smallest closest geometrical distance to the head block of the cooperative.
4. *Farthest Geometrically blocks*: in every iteration we add to the cooperative a block with the highest geometrical distance to the head block of the cooperative.
5. *Random*: Random set of blocks. The cooperative is built by uniformly adding blocks one by one, until the cooperative size is reached.

In our experiments, we run and compared various combinations of search algorithms with cooperative size and formation methods. The algorithms advance as follows:

once the cooperative has been formed, all the feasible permutations of possible moves for the cooperative (depending on the search algorithm) are calculated. For each permutation we calculate the individual cost of the cooperative in the resulting placement. The permutation that minimizes this cost is chosen. Only the cooperative cost is taken into account, and we ignore the global cost as well as the internal distribution of the cost among the blocks composing it.

2.5 Expected Algorithms' Progress

In this section we review our algorithms by describing their progress in general. A typical cost-over-time progress of BRD-ConstPerb is depicted in Fig. 5. Since players can choose a step which harms their individual cost, we expect the algorithm to be able to escape local minima by moving to a more expensive placement and improving it by a sequence of cost-reducing moves, which hopefully lead to a better local minimum.

The progress of our BRD-RlxSrch method depends heavily on the fines for overlaps. Recall that these fines increase with the run time. As illustrated in Fig. 6, this enables the algorithm to explore more areas in the solution space. Once the fines are above some threshold, the algorithm explores feasible or almost feasible solutions whose cost may be higher than former non-feasible solutions explored earlier.

Finally, Fig. 7 illustrates the typical cost-over-time progress of BRD-RlxSrch and BRD-ConstPerb when coordinated deviations are allowed. The possibility to accept worse or unfeasible solutions enables the algorithm to escape local minima and to advance in the search space towards a better solution.

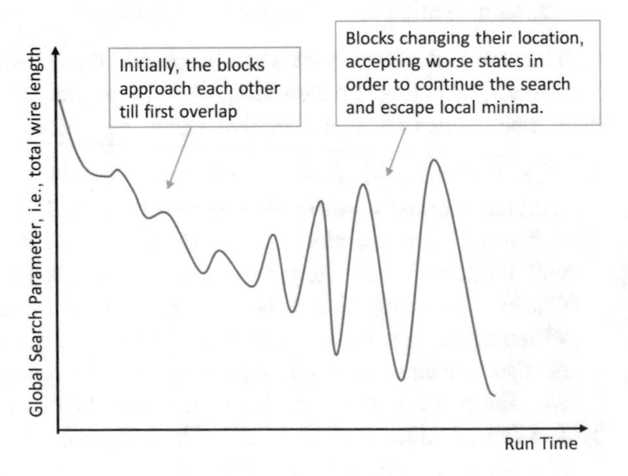

Fig. 5 Expected progress of the *BRD-ConstPerb* search method with unilateral deviations

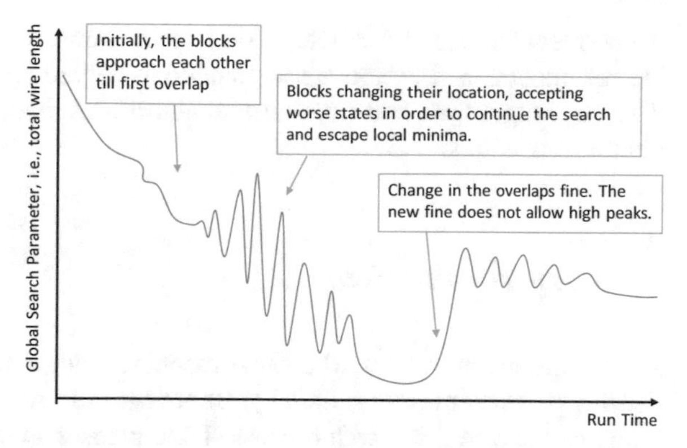

Fig. 6 Expected progress of the *BRD-RlxSrch* method with unilateral deviations. The slope has peaks, the cost function is monotonically decreasing via game of tradeoffs between the search parameters and the overlaps

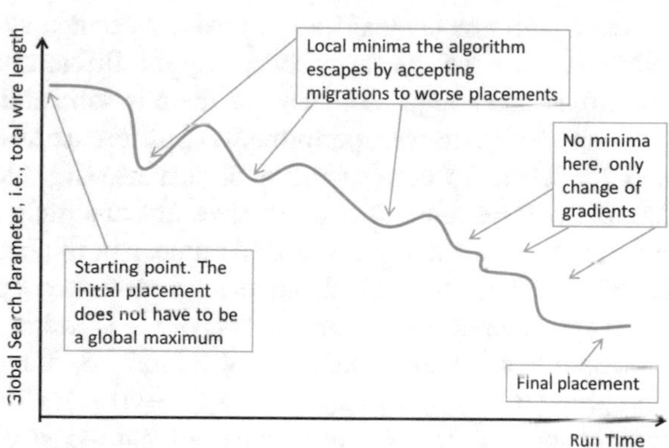

Fig. 7 Expected progress of *BRD-ConstPerb* and *BRD-RlxSrch* with coordinated deviation

In the above figures we present the slopes as monotonically nicely curved lines, in reality this is not the case. The real lines have various gradient changes, and they are far from being nicely curved over the monotonic movement sections. The figures present the tendency of the algorithm and the overall progress.

3 Experimental Results

In this section we present our experimental study. Our experiments simulate the global placement stage. This stage is followed by the detailed placement stage, in which the global placement results are applied and the blocks are actually placed on the die. Usually at the detailed placement stage, small changes occur in order to solve some local issues such as wire congestion spots, remaining overlaps, layout constraints, connecting to the die pinout, etc.

We first demonstrate our concept by presenting the results of the unilateral deviation algorithms. Next, we compare some of our heuristics with the *Simulated annealing* algorithm. Finally, we study coordinated deviations and analyze the effects of the cooperative size and structure. We explore how coordinated deviations improves the results obtained by unilateral deviations, regardless of the selected method for the search algorithm, and also consider algorithms that combine unilateral and coordinated deviations.

3.1 Experiments Setup

All algorithms are ran on the same machine with similar conditions. We sample various parameters during the algorithms run, in order to study not only the final outcome but also the search process. Time measurement is conducted and counted by the algorithms context timers, thus if a context switch occurs the timer pauses. While the time values themselves vary on different machines, the progress of the algorithms and comparison between them is valid and independent of the machine.

Our experiments were performed on 6 different test-benches, T_{30}^4, T_{30}^6, T_{30}^8, T_{40}^4, T_{40}^4 and T_{40}^8, where T_n^c corresponds to an instance of n blocks, with c connections-per-block. In all instances, the block sizes are randomly distributed, height and width being a random equally distributed number in the range [30, 80] (pixels). The different connections-per-block parameter enables a good comparison and allows us to isolate and emphasize various aspects of the algorithms.

Recall that the Individual cost of a block B_i in a placement F is defined to be $Ind\_cost(B_i, F) = \alpha \cdot (L_{B_i}(F))^2 + (1 - \alpha) \cdot A(F)$. We run the search algorithms with various values for α, and found out that the wire-length component should get much higher weight. Thus, all the results described in this section were obtained with $\alpha = 0.9$. This fits the common practice these days to focus on minimizing the wire-length of the placement and only when done optimizing the wire-length, perform small changes in order to gain better area result.

As detailed in Sect. 2.3, we used three local search method: BRD—only legal profitable moves, BRD-ConstPerb—legal but maybe harmful moves, and BRD-RlxSrch—profitable but maybe non-legal moves (overlaps associated with fine). These search methods are constructed into algorithms, combining unilateral players and coordinated deviation players.

Each of these local search methods run in two different variations, without or with *swap moves*. Note that a swap move differs from a cooperative of size 2. The two members of a cooperative may swap places as long as it improves their total cost. However, a swap move is initiated by a single block and may hurt significantly the individual cost of the second block involved.

In order to better evaluate the algorithms, we performed each experiment several times. Specifically, we ran the algorithms on the same instance with 5 different random initial placements. While the initial placement has a strong impact on the results, the final result depends on the progress of the algorithm as much as on the initial placement. If for any test bench one algorithm is better than the other, then it is almost certain better for any initial placement. The variation between the results of different algorithms is consistent for most of the initial placements. In order to compare the algorithms we look at the average results over all initial placements.

3.2 Results for Unilateral Deviations

The first experiment we present is a comparison of the three search methods, when applied without and with swap moves. We run each of these variants on our test-bench T_{40}^4, that is, an instance consisting of 40 blocks each with 4 connections. All the algorithms were applied starting from the same initial placement. Figures 8 and 9 present the progress of wire-length over running-time without and with swaps, respectively. Figures 10 and 11 present the progress of area over running time with and without swaps, respectively. We can see that the algorithms reach their stopping criteria at different times and have different progress gradient.

The BRD algorithm is the first to finish—its local search is more restricted and thus, the stopping criteria is reached earlier. Each algorithm has a different progress curve. The gradient of the changes in the cost function according to the time is

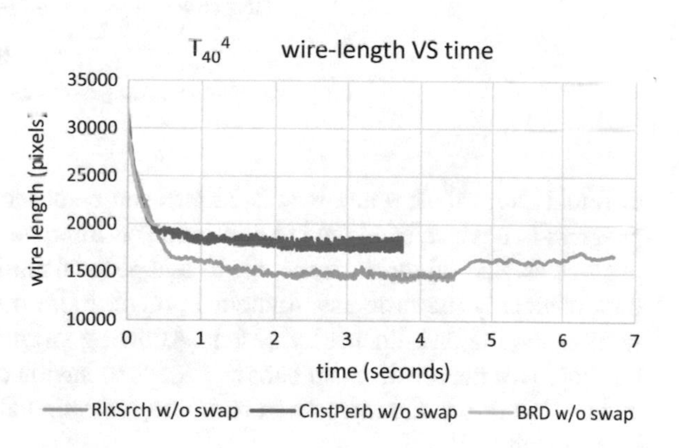

Fig. 8 Progress of total wire-length. No swap moves

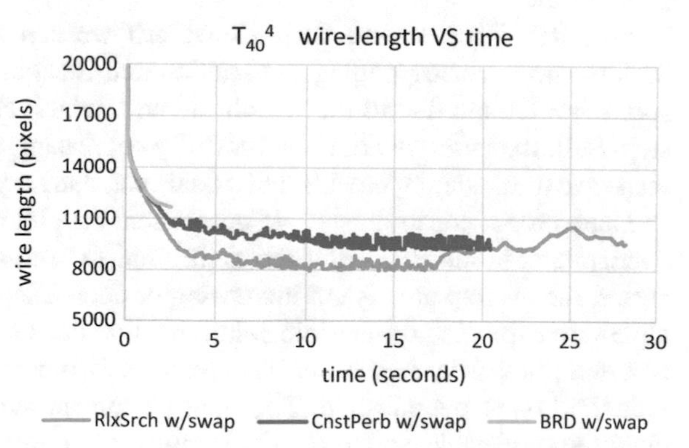

Fig. 9 Progress of total wire-length. Swap moves allowed

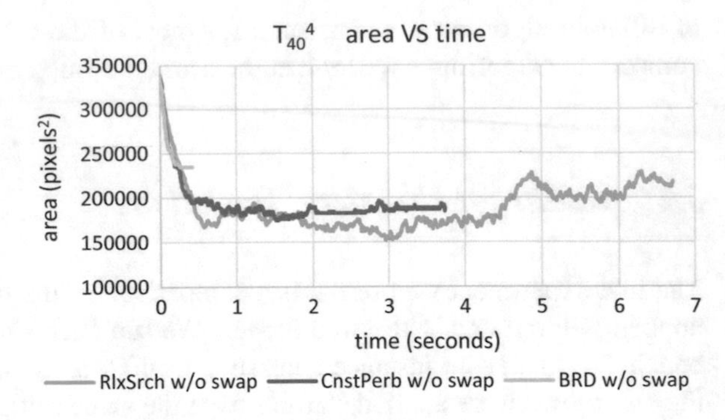

Fig. 10 Progress of bounding-box area. No swap moves

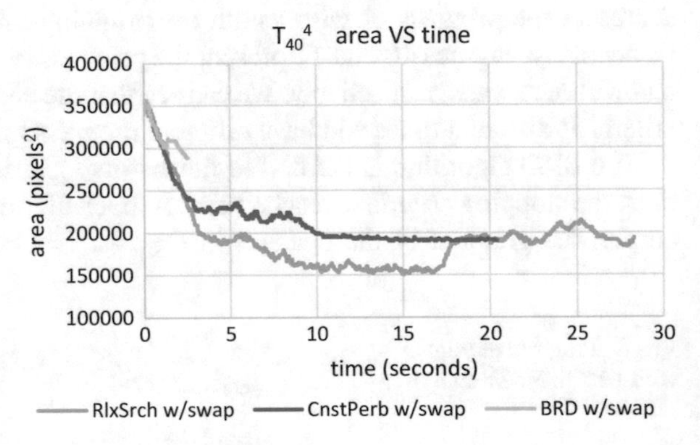

Fig. 11 Progress of bounding-box area. No swap moves

different. Nevertheless an obvious pattern can be observed: the algorithms that can progress to a worse state tend to continue and improve their result as the algorithm progresses. Such algorithms are able to escape local minima and continue the search, thus obviously the progress gradient is much more moderate. Moreover, we also witness the major influence of swaps. Allowing swaps increases the running time but improves the result. Such behavior leads to having a more moderate gradient of change, but due to the increase in run time, we eventually reach a lower level and a better result.

For BRD-RlxSrch and BRD-ConstPerb we can see peaks and drops in the performance—corresponding to reaching and escaping local minima. In BRD the spikes are limited due to the search method, which always chooses an improving step. Still, the curve is not monotonically decreasing as the improvement are with respect to the deviating block's individual cost, that may conflict with the global cost.

Due to space constraints we do not present the plot presenting the progress of the overlap area in the BRD-RlxSrch algorithm. In both applications, with or without swaps, the overlap area is not increasing or decreasing monotonically. Initially, the algorithm explores non-feasible solutions, that have low wire-length and bounding box area; however, as the fine for overlaps is increased, the placements become more and more overlap-free. The final placement achieved by the relaxed search algorithm is feasible, and its quality is more or less equivalent to the quality achieved by BRD-ConstPerb.

3.3 *Comparison with the* Simulated Annealing *Algorithm*

In this section we present a comparison of our unilateral deviation algorithms with the Simulated Annealing (SA) algorithm. We present the results by normalizing the SA result to 1. We ran SA and each of our algorithm on all six test-benches, and normalized the result with respect to the corresponding SA result. For example, if on some instance the SA algorithm produces a placement whose area is 12,000 pixel^2, and our algorithm produces a placement whose area is 9600 pixel^2, then the result of our algorithm is presented as $9600/12,000 = 0.8$.

When presenting the results, we distinguish between two groups of algorithms. The first group includes algorithms that are more run-time oriented than result-oriented, while the other aim to achieve a good result. The first group consists of BRD with and without swaps, and BRD-ConstPerb without swaps, while the second group consists of BRD-RlxSrch with and without swaps, and BRD-ConstPerb with swaps.

Figures 12 and 13 present the results for the total wire-length, and Figs. 14 and 15 present the results for the placement area. In all the figures, the horizontal black line represents the result achieved by the SA algorithm.

Figures 16 and 17 present the comparison between the run-times of the algorithms. The algorithm of the first group are indeed much faster than SA. Also, all algorithms when run without swaps are at least 5 times faster than SA.

The experiments reveal that we can achieve better results of both wire-length and placement area while paying with a slightly worse running time. As well as the other way around, that is, slightly worse result can be achieved with a fraction of the running time. We also witness certain algorithms, which on the majority of the test benches, have succeeded to achieve a better result in a lower running time.

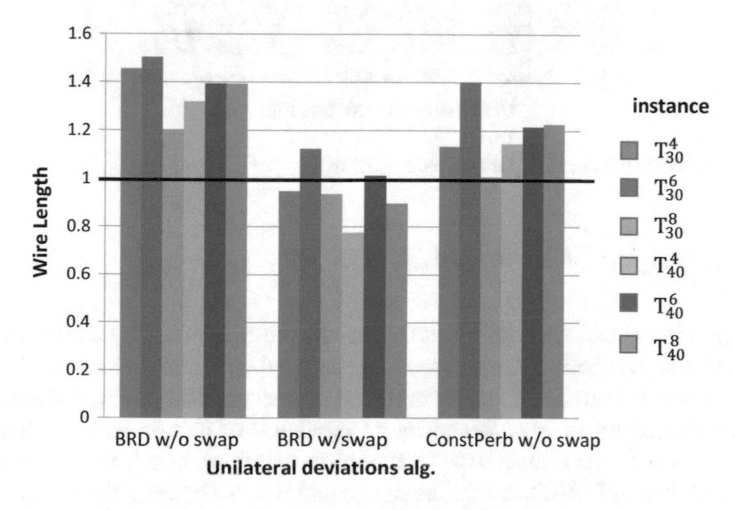

Fig. 12 Wire-length normalized to SA performance—group I

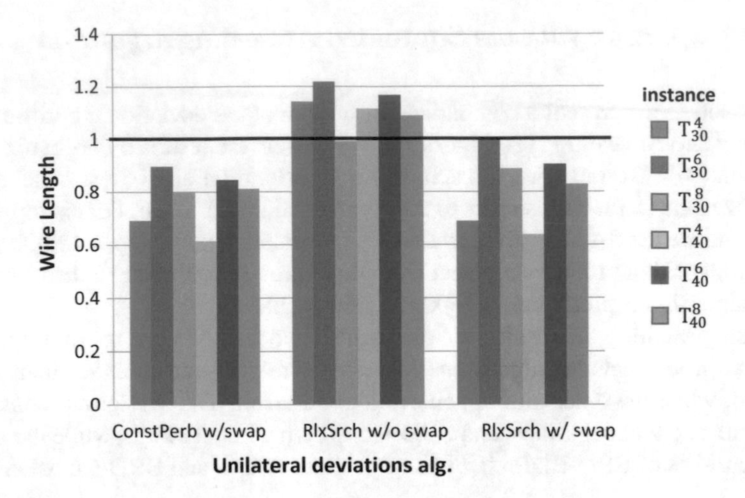

Fig. 13 Wire-length normalized to SA performance—group II

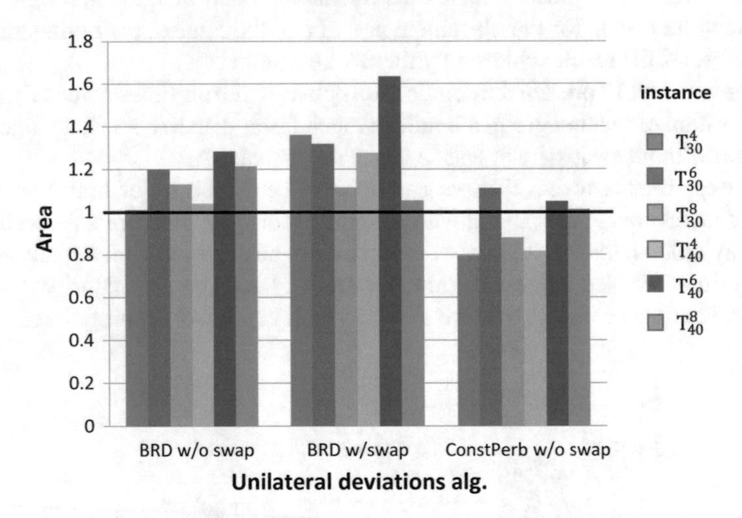

Fig. 14 Area normalized to SA performance—group I

3.4 Results for Coordinated Deviations

As detailed in Sect. 2.4, coordinated deviations are performed by a cooperative of blocks. All the members of the cooperative change their strategy (location) together. In this section we analyze the results achieved by our search algorithms when applied with coordinated deviations. Recall that a deviation of a cooperative is beneficial if the total cost of the cooperative members is strictly reduced. In addition to the local-search method (BRD, BRD-ConstPerb and BRD-RlxSrch), the algorithms are different in the way they determine the active cooperative size and formation. In all

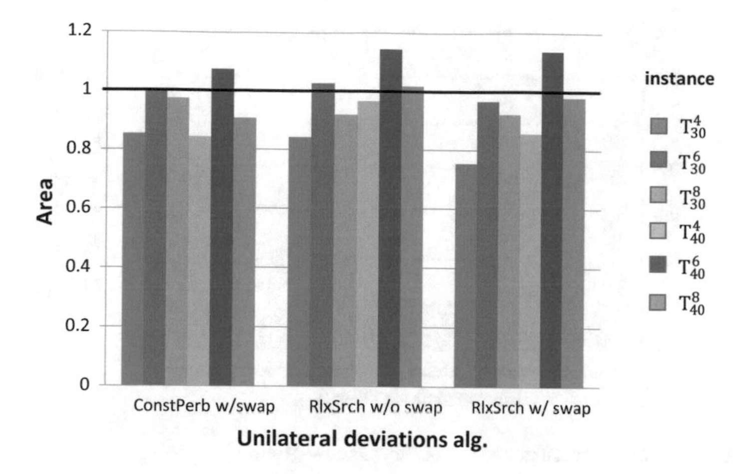

Fig. 15 Area normalized to SA performance—group II

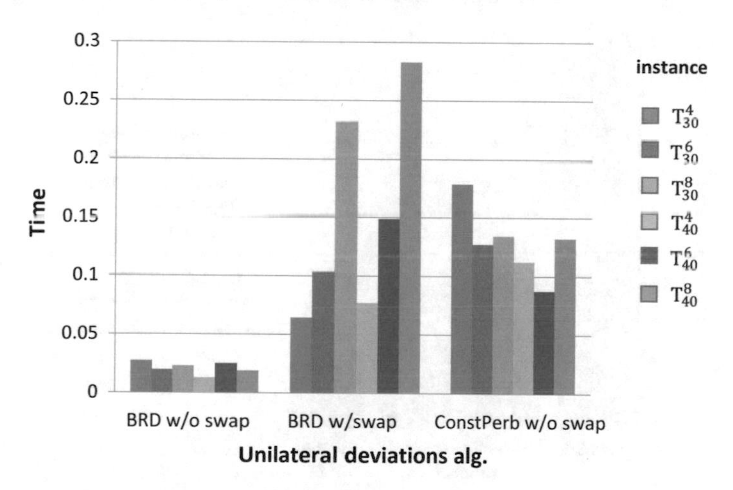

Fig. 16 Running time normalized to SA performance—group I. The line corresponding to SA (Time = 1) is not shown, as it is way above the shown bars

the experiments with coordinated deviations, the algorithms were performed on the same instance and the same initial placements.

A sample of our results are presented in Figs. 18 and 19. Our results show that the cooperative size has the largest influence on the results. The method for determining the cooperative size in each round is not crucial as the predefined limit for the maximal cooperative size. The higher this limit, the better the results. In addition, the experiments do not crown a specific method for selecting the cooperative members—the results vary and depend heavily on the initial placement. The results in the figures are present in comparison to Simulated Anneling (SA). That is, we run SA on the same instance and the same initial placement, and the peformance of our algorithms

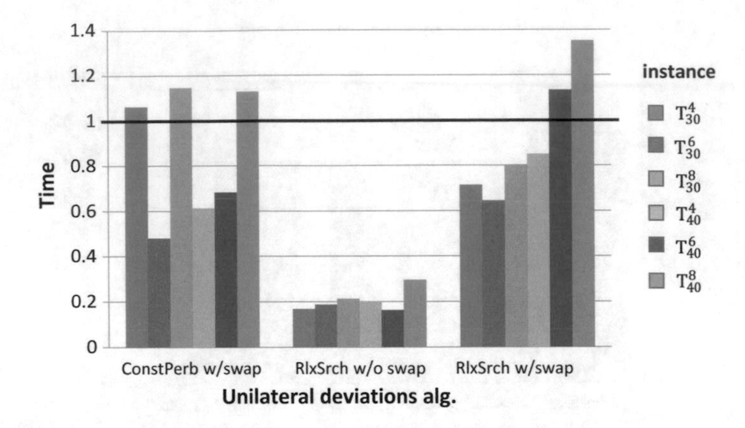

Fig. 17 Running time normalized to SA performance—group II

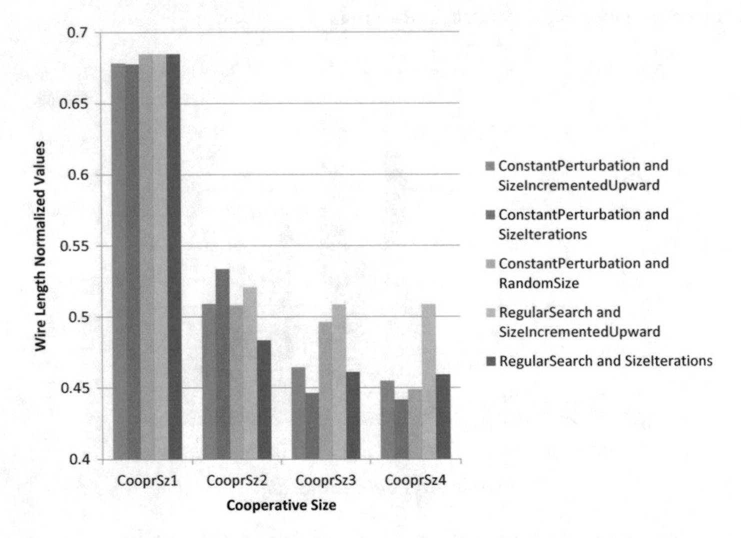

Fig. 18 Wire-length results for different cooperative size. In this experiment, the cooperatives are formed using the *Closest Geometrically block* method

is shown as a fraction scaled according to SA. It can be seen that our algorithms constantly achieve better results.

The running time is presented in Fig. 20. We can see that the running time increases exponentially with the cooperative size. This does not come as a surprise as we know the problem is NP-hard.

The best results were achieved when iterating over different cooperative size, where the size increases after each round till a predefined limit. We witnessed a major improvement in the results already with cooperatives of two blocks (compared with unilateral deviations). Further increase in the cooperative size do tend to improve the result, however it involves an exponential increase in the running time. Therefore,

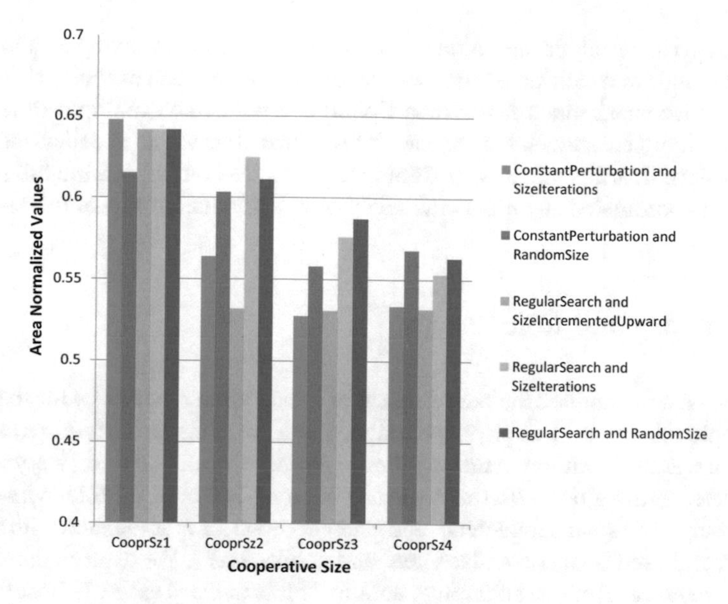

Fig. 19 Area results for different cooperative size. In this experiment, the cooperatives are formed using the *Farthest Connected block* method

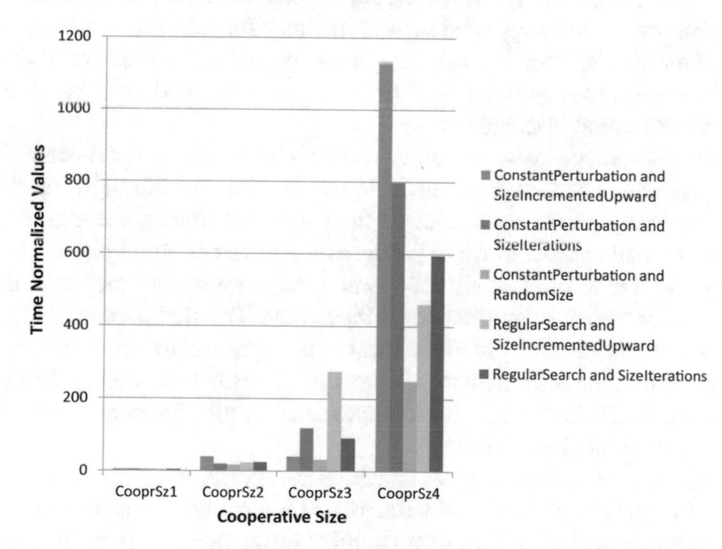

Fig. 20 Running time for different cooperative size. In this experiment, the cooperatives are formed using the *Closest Connected block* method

the best is to run the algorithm initially with cooperatives of size two, and allow non-frequent rounds in which larger cooperatives are activated. Such executions converge to a final placement much faster than SA, and if performed multiple times, with different initial placements, are expected to produce at least one excellent outcome.

We believe that this unique algorithm, that combines our search methods with a mixture of coordinated and unilateral deviations, is the main result of this work.

4 Summary and Conclusions

In this work we examined the performance of local search algorithms for the global placement problem in VLSI physical design. Our algorithms are different from common local search algorithms in the way they explore the solution space. Every solution is associated with a global cost (based on its bounding-box area, the total wire-length, and possibly additional parameters), and every component is associated with its individual cost (based on its own placement and connections). We explore the solution space by moving from one solution to another if this move is selfishly beneficial for a single or for a cooperative of components, without considering the effect on the global cost. Best-response dynamics (BRD) is performed until no component has a beneficial migration. We suggested several methods for selecting the component(s) initiating the next step, and for selecting their migration. In order to evaluate our algorithms, we have tested them on various test-benches, and each test-bench was ran with various initial placements.

Based on our experiments, we can distinguish between two approaches for handling the problem. The first approach is to use algorithms with high run-time that also tend to supply good results. Due to their high run-time, these algorithms can only be ran a small number of times (with several different initial placements). The second approach is to use fast algorithms and ran them many times with the hope to get at least one good output in one of these runs. The first approach rely on the algorithms' ability to consider multiple local minima, thanks to their ability to escape local minimum. In the second approach the algorithms tend to stop at the first local minimum they found, however, this is compensated by the high number of runs, with many different initial placements.

Our algorithms, even for instances on which they perform relatively poor, achieve results not far from SA with only a fraction of its running time. This feature obviously can be very handy when one tries to get a quick estimation of the results achievable for a given instance. We believe that this work has proved the concept of selfish local search to be valid and efficient. Moreover, this concept may be useful in solving additional optimization problems arising in real-life applications.

References

1. Adya, S.N., Markov, I.L.: Combinatorial techniques for mixed-Size placement. ACM Trans. Des. Autom. Electron. Syst. **10**(1) (2005). https://doi.org/10.1145/1044111.1044116
2. Alpert, C.J., Mehta, D.P., Sapatnekar, S.S.: Handbook of Algorithms for Physical Design Automation. Auerbach Publications (2008)
3. Chang, Y.C., Chang, Y.W., Wu, G.M., Wu, S.W.: B*-trees: a new representation for non-slicing floorplans. In: Proceedings of the 37th Annual Design Automation Conference, pp. 458–463 (2000). https://doi.org/10.1145/337292.337541
4. Chu, C.:. Electronic Design Automation: Synthesis, Verification, and Test, Chapter 11: Placement, pp. 635–685. Springer (2009)
5. Even-Dar, E., Mansour, Y.: Fast convergence of selfish rerouting. In: Proceedings of the 16th Annual ACM-SIAM Symposium on Discrete Algorithms (SODA), pp. 772–781 (2005)
6. Feldman, M., Snappir, Y., Tamir, T.: The efficiency of best-response dynamics. In: Proceedings of the the 10th International Symposium on Algorithmic Game Theory (SAGT) (2017)
7. Gerez, S.H.: Algorithms for VLSI Design Automation. Wiley, New York (1999)
8. Ghosal, P., Samanta, T.: Symposium on VLSI Thermal-Aware Placement of Standard Cells and Gate Arrays: Studies and Observations. IEEE Computer Society Annual (2008). https://doi.org/10.1109/ISVLSI.2008.37
9. Huang, E., Korf, R.E.: Optimal rectangle packing: an absolute placement approach. J. Artif. Intell. Res. **46**, 47–87 (2012)
10. Intel Core *i*7 processors (2008). www.intel.com/content/www/us/en/products/processors/core/i7-processors.html
11. Jiang, Z., Chen, H., Chen, T., Chang, Y.: Challenges and solutions in modern VLSI placement. In: International Symposium on VLSI Design, Automation and Test (VLSI-DAT) (2007)
12. Kleinberg, J., Tardos, E.: Chapter 12: Local Search. Algorithm Design, pp. 690–700. Addison-Wesley (2005)
13. Khachaturyan, A., Semenovskaya, S., Vainshtein, B.: Statistical-Thermodynamic Approach to Determination of Structure Amplitude Phases. Sov. Phys. Crystallography. **24**(5), 519–524 (1979)
14. Kahng, A.B., Lienig, J., Markov, I.L., Hu, J.: VLSI Physical Design: From Graph Partitioning to Timing Closure. Springer (2011). https://doi.org/10.1007/978-90-481-9591-6
15. Lengauer, T.: Combinatorial Algorithms for Integrated Circuits. Wiley-Teubner (1990). https://doi.org/10.1007/978-3-322-92106-2
16. Lin, Y., Yu, B., Xu, X., Gao, J., Viswanathan, N., Liu, W., Li, Z., Alpert, C.J., Pan, D.Z.: MrDP: multiple-row detailed placement of heterogeneous-sized cells for advanced nodes. IEEE Trans. Comput.-Aided Des. Integr. Circ. Syst. **37**(6), 1237–1250 (2018)
17. Markov, I.L., Hu, J., Kim, M.: Progress and challenges in VLSI placement research. Proc. IEEE **103**(11), 1985–2003 (2015)
18. Murata, H., Kuh, E.S.: Sequence-pair based placement method for hard/soft/pre-placed modules. In: Proceedings of the International Symposium on Physical Design, pp. 167–172 (1998)
19. Norman, R., Last, J., Haas, I.: Solid-state micrologic elements. In: Solid-State Circuits Conference. Digest of Technical Papers, pp. 82–83 (1960). https://doi.org/10.1109/ISSCC.1960.1157264
20. Pattanaik, S., Bhoi, S.P., Mohanty, R.: Simulated annealing based placement algorithms and research challenges. J. Global Res. Comput. Sci. **3**(6) (2012)
21. Quinn, N., Breuer, M.: A forced directed component placement procedure for printed circuit boards. IEEE Trans. Circ. Syst. **26**(6) (1979)

22. Reda, S., Chowdhary, A.: Effective linear programming based placement methods. In: Proceedings of the International Symposium on Physical Design, pp. 186–191 (2006)
23. Rutenbar, R.A.: Simulated annealing algorithms: an overview. IEEE Circ. Dev. Mag. **5**(1), 19–26 (1989)
24. Shahookar, K., Mazumder, P.: VLSI cell placement techniques. ACM Comput. Surv. **23**(2), 143–220 (1991)
25. Yang, Z., Areibi, S.: Global placement techniques for VLSI physical design automation. In: Proceedings of the 15th International Conference on Computer Applications in Industry and Engineering (2002)

Application of Information Systems and Technologies in Transport

Kristina Pavlova and Vladimir Ivanov

Abstract The use of information systems in transport is crucial for the construction, maintenance, formation, management and use of different types of transport services. They make it possible to achieve a comprehensive coverage of the different modes of transport, based on current regulatory and legal requirements in terms of technology, and financial considerations. This opens the possibility to manage various activities related to the optimization of travel and their quality, as well as to collect data in order to optimize, manage and minimize transport costs in road, water, air and rail transport systems. In this sense, an information system is one that, in the form of interconnected components, covers all forms of collecting, storing, retrieving, processing and providing information about the organization concerned. For this reason, the tendency in their development and implementation determines an aspiration for presentation at a new higher level capable of providing increased efficiency, precision and validity of management decisions. In this respect, the present paper examines the creation of a mathematical model allowing the intensification of rail passenger transport by introducing their baseline indicators in an information system, which is a basic trend for European countries. This model was used to create an information system allowing to solve a two-hierarchical optimization problem for train management between Sofia and Varna.

Keyword Transport · Mathematical model · Optimization

1 Introduction

In essence, the expression transport system is a concept embracing the aggregate of all modes of transport linked in accordance with applicable normative and legal

K. Pavlova · V. Ivanov (✉)
Institute of Information and Communication Technologies, Acad. G. Bonchev Bl 2, Sofia, Bulgaria
e-mail: Ivanov.vladi@gmail.com

K. Pavlova
e-mail: kristina@hsi.iccs.bas.bg

S. Fidanova (ed.), *Recent Advances in Computational Optimization*, Studies in Computational Intelligence 920,
https://doi.org/10.1007/978-3-030-58884-7_9

requirements in respect of technic, technologies and financial considerations. For many areas related to the construction, maintenance and management of deferent types of transport systems, the formation and use of services related to the use of information systems in transport is of paramount importance. Through them, the transport companies are able to manage different aspects of their activity related to improving the quality of travel and solving travel optimization tasks as well as collecting data allowing for processing, revealing new methods for optimization of management and minimization of transport costs [1, 2]. There are different types of transport systems among which automobile, water, air and rail are predominant. Automobile transport systems offer a relatively high speed of picking and delivering loads in rails where there are no other modes of transport. The water transport systems provide mass transport to ports within and outside the country. Their cost is much lower than other types, especially when transporting long distances. The air transport systems are characterized as the most expensive and fast commuters of people and goods at short and long distances. Rail transport systems occupy a foothold in many industrialized countries. They use modern locomotives and wagons, a powerful track, modern automation, telemetry and computing equipment. This determines their universality and opportunity to serve the country's manufacturing industries and to satisfy the needs of the population regardless of climatic conditions.

2 Essence of an Information System

Generally, the information system is a means of encompassing all forms of collecting, storing, retrieving, processing and disseminating information. It represents a multitude of interrelated components and data about people, places and objects that are relevant to the organization. On this basis, it is capable of delivering information services, supporting the decision-making process, which closes the circle of all kinds of information activities that have a place in the organization that uses it to analyze problems and create new products.

The current trends in development and supply and information systems are characterized by the continuous increase of their capabilities and specialization to the requirements of the users working in the sphere of management. This allows each next generation of information systems to be presented at a new high level capable of ensuring the technical and technological attitude of management processes, thereby increasing the efficiency, precision and soundness of managerial decisions [1, 2].

3 Information System for Solving a Two Level Hierarchical Tasks

In order to optimize a transport network, it is necessary to create a mathematical model, numerically identify the potential areas where rail passenger transport can be intensives; this can be achieved by entering the available data into an information system and solving a task for and a maximum flow. If no such program is created, it is programmed for the respective task. As a result, solving such a problem will allow for the increase in passenger traffic carried by rail passenger transport, which tends to increase the use of rail transport in Europe [3, 4]. The information system implements and evaluates the hierarchical optimization model as:

- Determines a defined two-hierarchical optimization control task.
- Determines the magnitude of the maximum ow between Sofia and Varna.
- Defines a flow distribution giving priority to train traffic.

The task of “maximum flow” can be considered as a task for network system analysis. The task of analyzing a transport network can be defined as, for a given network topology with network connection capacity values, it is necessary to determine the maximum amount of flow between two nodes that can be transmitted over the network. The analytical formulation of the maximum flow task is presented in the form (2)

$$Max_{\substack{i \in A(j) \\ j \in B(i) \\ i \neq j = 1, N}} \; [f_{st}] \tag{1}$$

$$\sum_{j \in A(i)} f_{ij} - \sum_{j \in B(i)} f_{ji} = f(x) = \begin{cases} 0, & i \neq s, t \\ f_{st}, & i = j \\ -f_{st}, & i = t \end{cases} \tag{2}$$

$$0 \leq f_{st} \leq v_{ij} \forall_{i,j} \in 1, N \tag{3}$$

where i, j represents the number of nodes in a network with N nodes, A(i) = j in N set of nodes i, which initiates the inbound links to the node in j, B(i) = j N set of nodes i, which are connected to the output link from node j, s and t denote the start and end nodes between which the maximum flow that can be transmitted through the network is sought; v i, j the throughput of the arcs between nodes i and j, (i, j) N; f(i, j) the unknown streams that need to be evaluated as components of the maximum network flow.

Dependency (2) describes the law of continuity of flows. The amount of incoming stream is equal to the amount of outgoing stream from the same node. The dependence (2) of the start and end nodes reflects that a stream quantity is output at the starting node, and the end node t enters the same amount of flow, which is reflected by a minus sign [5]. Dependence (3) shows a physical limitation that the flow of a given arc can not be greater than the throughput of that arc. The target function of the optimization task (1) formalizes the flow maximization requirement that starts from the initial node s and enters the end point t [6, 7].

Task solutions (1) determine the value of the maximum flow that can be transmitted between the network from s to t. In addition, the maximum flow components also determine the paths through which the individual components of the maximum flow across the entire network pass. Therefore, at a given network arc capacity, the maximum flow task (1) gives the amount of flow that can pass through the predefined network topology and the paths of the individual components that make up the maximum flow [8]. Tasks with a network structure can be defined besides analysis tasks and a synthesis task. In order for a unit stream to flow through an arc, it is necessary to pay a certain resource/price. Such a task for optimal flow distribution synthesis is as follows: to define such a flow distribution in the network, whereby the value of the total flow distribution is the least. This synthesis task determines such a flow distribution across the network, whereby a given volume of traffic is transmitted in directions in order to realize the smallest value of the flow distribution. This synthesis task is used to target the components of the maximum flow of arcs that are supported by rail transport. Thus, by setting a low value of the flow distribution over the arc of the railway transport and a high value of flow distribution over the arcs supported by bus transport, maximum utilization of the railway transport capacity will be realized [9]. The use of these two optimization tasks for analysis and synthesis will allow a common mathematical model to define such a distribution of passenger transport as a solution whereby rail transport will be of paramount importance and will intensify its operation. Linking both analysis and synthesis tasks will be performed in the following sequence. The task of analysis will determine the value of the maximum flow that can be transmitted to the transport network in the direction Sofia-Varna. Limitations shall take account of the throughput of the arcs determined by bus and rail timetables on the journey time estimate for a given segment [10]. The synthesis task will use the value of the specified maximum stream as a constraint for the amount of traffic required. As a result, a solution for the lowest-cost flow distribution will be used and will benefit from rail transport. The resulting flow distribution from the synthesis task will define new pass through capabilities for the analysis task. Thus, the ideology of this new mathematical model consists in defining and solving a two-hierarchical optimization task that will provide both a priority distribution of the use of rail transport as well as an automatic assessment to reduce the intensity of bus transport.

4 Solution of Optimization Task for Maximum Flow

The analytical view of the task of finding the maximum flow between two nodes (city) in a transport scheme is of the type (1). The defined optimization task in analytical form as "maximum flow task" is the type of linear programming. As constraints of the optimization task, the requirements for preserving the flow passing through a given node (1b) are defined. For the start node Sofia, which is marked with the number (s) and the end node - Varna, which is marked with the number (t), the flow-saving equations contain the variable with unknown value for the maximum flow f_{st}. An information system has been set up to determine the flow consistency equations. They are defined on the principle of the flow that enters is equal to the flow that flows out. The flow-saving equation for the initial node Sofia, s = 1 is:

$$f_{1,2} + f_{1,3} + f_{1,4} + f_{1,5} + f_{1,9} + f_{1,11} + f_{1,15} + f_{1,13} - f_{st} = 0 \quad (4)$$

where $f_{i,j}$ is the flow between nodes i and j. the flow-keeping equation for node 2 is:

$$f_{1,2} - f_{2,3} = 0 \quad (5)$$

the flow consistency equation for node 3 is:

$$f_{2,3} - f_{3,4} + f_{1,3} = 0 \quad (6)$$

the flow consistency equation for node 4 is:

$$f_{3,4} + f_{1,4} - f_{4,15} - f_{4,13} - f_{4,9} - f_{4,5} = 0 \quad (7)$$

the flow consistency equation for node 5 is the dependence:

$$f_{4,5} + f_{1,5} - f_{5,6} = 0 \quad (8)$$

the flow consistency equation for node 6 is:

$$f_{5,6} - f_{6,7} = 0 \quad (9)$$

the flow consistency equation for node 7 is:

$$f_{6,7} + f_{15,7} - f_{7,8} - f_{7,8} - f_{7,9} = 0 \quad (10)$$

the flow consistency equation for node 8 is:

$$f_{7,8} - f_{8,9} = 0 \quad (11)$$

the flow consistency equation for node 9 is:

$$f_{7,9} + f_{8,9} + f_{4,9} + f_{1,9} + f_{15,9} - f_{9,10} - f_{9,11} = 0 \quad (12)$$

the flow consistency equation for node 10 is:

$$f_{9,10} - f_{10,11} = 0 \quad (13)$$

the flow consistency equation for node 11 is:

$$f_{10,11} + f_{9,11} + f_{1,11} + f_{15,11} + f_{16,11} - f_{11,12} - f_{11,13} - f_{11,17} = 0 \quad (14)$$

the flow consistency equation for node 12 is:

$$f_{11,12} - f_{12,13} = 0 \quad (15)$$

the flow consistency equation for node 13 is:

$$f_{11,13} + f_{12,13} + f_{4,13} + f_{1,13} + f_{15,13} + f_{16,13} - f_{13,17} - f_{13,14} = 0 \quad (16)$$

the flow consistency equation for node 14 is:

$$f_{13,14} - f_{14,17} = 0 \quad (17)$$

the flow consistency equation for node 15 is:

$$f_{1,15} + f_{4,15} - f_{15,7} - f_{15,9} - f_{15,11} - f_{15,13} - f_{15,17} - f_{15,16} = 0 \quad (18)$$

the flow consistency equation for node 16 is:

$$f_{1,16} + f_{15,16} - f_{16,11} - f_{16,13} - f_{16,17} = 0 \quad (19)$$

the equation for preserving the flow in the final node Varna, t = 17 is:

$$f_{14,17} + f_{13,17} + f_{11,17} + f_{15,17} + f_{16,17} - f_{st} = 0 \quad (20)$$

The flows in a given arc of the transport scheme must take into account the arc crossing limits in the scheme: The goal of the maximum flow task is to maximize the value.

$$0 \leq f_{st} \leq v_{ij} \forall_{i,j} \in 1, N \quad (21)$$

A defined maximum flow assignment is numerically defined. It contains 44 variables and 61 limitations in total. Of these limitations, 17 account for the continuity of the flow in the 17 nodes (city) of the transport scheme and 44 constraints are to respect the throughput of the arcs in the transport scheme. The solution for the "maximum flow" task was done using the MATLAB information system. In its default configuration, this software does not contain any means to solve the "maximum flow" task. An additional programming module is added, which is an add-on integrated into the MATLAB environment, which can solve the required task [11]. The add-in consists of a new function called graphmaxflow [12]. This function takes input parameters and calculates the value of the maximum flow and the arcs through which the maximum flow components pass [13, 14].

The graphmaxflow (G, S, T) function calculates the maximum flow in a given scheme G from node S to node T. G is the n-of-n matrix that represents the scheme. Non-zero records in G determine the capacity of the ribs. M is the maximum flow, $F_{(i,j)}$ is the flow from node i to j. The flow matrix parameter displays the capacity value of each arc of bus and rail traffic. The resulting solutions are presented in tabular form. The denominator of the fraction is the calculated throughput of this arc according to the developed algorithm for data preparation. The Liquid Numerator is the solution for the maximum flow task.

This number shows how much of this arc is being used by the maximum flow components when they move from the start to the end point [10].

Table 1 shows the results of the assigned task. The first number is the results obtained, the permeability of the arcs in the scheme, and the second number is the maximum flow in relative units. If the throughput of the arcs is equal to the maximum flow, then the capacity of this arc is filled and can not take up additional shipments. If the bandwidth is less than the maximum flow, then this arc may take up additional shipments.

Example: On the railway line, for the arc Targovishte-Shumen the throughput is 0.0623 and the maximum flow is 0.1305. This line may take on additional shipments. This can be done by reducing bus transport and transferring part of the passenger flow from buses to rail. The data used for the loading of the transport system is data for the year 2018. From the optimization tasks solution, it follows that a set of links supported by rail transport have a capacity that is not currently in use. This makes it possible to intensify the use of rail transport by redistributing the flows of the maximum flow so that they advantageously pass over the arcs supported by

Table 1 Results for the arcs in the diagram

Route trains	Route trains
Pleven-Levsk 0.0558/0.123	(Sofia-Pleven) 0.0861/0.0861
(Levski-Gorna Oryahovitsa) 0.0948/0.0948	(Sofia-Levski) 0.0390/0.04
(G. Oryahovitsa-Popovo) 0.0948/0.1037	(Shumen-Varna) 0.2202/0.2202
(Popovo-Targovishte) 0.0993/0.1294	(Sofia-Popovo) 0.0031/0.0031
Targovishte-Shoumen) 0.0623/0.1305	(Pleven-Popovo) 0.0186/0.0186
(Shumen-Varna) 0.0188/0.0413	(Sofia-Targovishte) 0.0062/0.0062
(Strazhitsa-Popovo) 0.0948/0.1061	(Popovo-Targovishte) 0.0172/0.0172
	(Veliko Tarnovo-Targovishte) 0.0917/0.1074
	(Antonovo-Targovishte) 0.0014/0.0443
	(Pleven-Shumen) 0.0054/0.0054
	(Targovishte-Shumen) 0.1242/0.2337
	(V.Tarnovo-Shumen) 0.0258/0.0258
	(Antonovo-Shumen) 0.0190/0.0190
	(Sofia-V.Tarnovo) 0.15570.1557
	(Pleven-V.Tarnovo) 0.0063/0.0063
	(Sofia-Antonov) 0.0036/0.0036
	(Veliko Tarnovo-Antonov) 0.0289/0.1364

rail. Which components of the maximum flow should be redirected from bus to rail is desirable to identify through the use of information technology rather than by expert opinion—by choosing a decision maker. This will result in an optimal flow distribution of traffic with a priority for the use of rail transport.

5 Conclusion

Information systems contribute to a number of advantages in transport management:

- Increasing the use of desired transport in a transport network.
- Transports costs are reduced - thanks to the ability to analyze selected routes.
- Objects in the transport network are used more rationally.
- Traffic data is transparent.
- The quality of logistics services is generally improved;
- Weak spots in the transport network can be identified and steps taken to remove them.
- Traffic on a transport network can be reduced.
- Determining the optimal flow distribution of traffic in a transport network. Determines a defined two-hierarchical optimization control task.

Using transport information systems can improve the quality of the trip, reduce traffic, increase the costs of transport companies, determine the quantity of lines in a given direction, collect ticket sales data, and more. Information systems are a necessary part of our daily lives. They have gone wide in all areas. The application of this kind of systems is another step forward in order to achieve more optimal management in the transport sphere.

6 Conclusion

The research has been carried out under the Scientific Research Fund under. The "Contest for the Financing of Scientific Research of Young Scientists and Post-doctoral Students-2018, Contract KP-06-M27/9 of 2018" "Contemporary Digital methods and tools for exploring and modeling transport flows".

References

1. Agarwal, R., Lucas, H.: The information systems identity crisis: focusing on high visibility and high-impact research. MIS Q. J. **381**, 398 (2005)
2. Mccluskey, T.L., Kotsialos, A., Mller, J.P., Klgl, F., Rana, O., Schumann, R.: Autonomic road transport support systems, p. 304. Birkhuser (2012)
3. Gegov, E., Postorino, M., Gegov, A., Vatchova, B.: Space independent community detection in airport networks. Complex Systems. Relationships between Control, Communications and Computing. Studies in Systems , pp. 211–248 (2016)
4. Gomory, R.E., Hu, T.C.: Multi-terminal network flows. J. Soc. Ind. Appl. Math. SL **5**, 551–570 (1961)
5. Knzi, H.P., Krelle, W.: Nonlinear Programming, vol. 5, p. 303. Berlin (1962)
6. Mansour, O., Ghazawneh, A.: Research in information systems: implications of the constant changing nature of IT capabilities in the social computing era. In: Proceedings of the 32nd Information Systems Research Seminar in Scandinavia (2009)
7. Vatchova, B.E., Pavlova, K.T., Paunova, E.N., Stoilova, K.P., Author, F.: Deep Learning Of Complex Interconnected Processes For Bi-Level Optimization Problem Under Uncertainty. Published by Scientific Technical Union of Mechanical Engineering Industry, vol. 4, pp. 18–19 (2018)

8. Fang, S., Guo, P., Li, M., Zhang, L.: Bi level multi objective programming applied to water resources allocation. Math. Problems Eng. 9–15 (2013)
9. Pavlova, K., Stoilov, T.: Mathematical model for increasing the efficiency of passenger railway transport in Bulgaria. In: X International Conference for Young Researchers, vol. 5, pp. 10–13. Technical Science Industrial Management (2016)
10. Pavlova, K., Stoilov, T., Stoilova, K.: Bi-level model for public rail transportation under incomplete data. Cybern. Inf. Technol. **5**, 199–110 (2017)
11. Trichkova, E., Trichkov, K.: Technological solution for automating and managing of business processes. In: Proceedings of UNITE Doctoral Symposium on Future Internet Enterprise Systems, vol. 5, pp. 16–21 (2011)
12. https://www.mathworks.com/help/bioinfo/ref/graphmaxflow.html. Last accessed 21 Nov 2016
13. Camacho-Vallejo, J.F., Cordero-Franco, E., Gonzlez-Ramrez, R.G.: Solving the bilevel facility location problem under preferences by a Stackelberg-evolutionary algorithm. Math. Problems Eng. **2014**, 14–20 (2014)
14. Ivanov, V.l.: The characteristics of transport traffic measuring. XXV International Scientific Conference. Trans. Mot. Auto. **5**, 112–115 (2017)

Online Algorithms for 1-Space Bounded Cube Packing and 2-Space Bounded Hypercube Packing

Łukasz Zielonka

Abstract Two online packing algorithms are presented: 1-space bounded cube packing algorithm with asymptotic competitive ratio 10.872 and 2-space bounded cube and d-dimensional hypercube packing method with asymptotic competitive ratio $(3/2)^d + 2$ for $d \geq 3$.

Keywords Online algorithms · Bin packing · Cube · One-space bounded · Two-space bounded

1 Introduction

In the bin packing problem, we receive a sequence of items of different sizes that must be packed into a finite number of bins in a way that minimizes the number of bins used. When all the items are accessible, the packing method is called *offline*. The packing method is called *online*, when items arrive one by one and each item has to be packed irrevocably into a bin before the next item is presented.

In the online version of packing a crucial parameter is the number of bins available for packing, i.e., *active bins*. Each incoming item is packed into one of the active bins; the remaining bins are not available at this moment. If we close one of the current active bins, we open a new active bin. Once an active bin has been closed, it can never become active again. When the method allows at most t active bins at the same time, it is called *t-space bounded*. Unbounded space model does not impose any limits on the number of active bins. It is natural to expect a packing method to be less efficient with fewer number of active bins. In this paper, we study both 1-space bounded cube packing and 2-space bounded hypercube packing.

Let S be a sequence of items. Denote by $A(S)$ the number of bins used by the algorithm A to pack items from S. Furthermore, denote by $OPT(S)$ the minimum

Ł. Zielonka (✉)
Institute of Mathematics and Physics, UTP University of Science and Technology,
Al. Prof. S. Kaliskiego 7, 85-789 Bydgoszcz, Poland
e-mail: Lukasz.Zielonka@utp.edu.pl

S. Fidanova (ed.), *Recent Advances in Computational Optimization*,
Studies in Computational Intelligence 920,
https://doi.org/10.1007/978-3-030-58884-7_10

possible number of bins used to pack items from S by the optimal offline algorithm. By the asymptotic competitive ratio for the algorithm A we mean:

$$R_A^\infty = \limsup_{n\to\infty} \sup_S \left\{ \frac{A(S)}{OPT(S)} \mid OPT(S) = n \right\}.$$

1.1 Related Work

The one-dimensional case of the space bounded bin packing problem has been extensively studied and the best possible algorithms are known: the Next-Fit algorithm [8] for the one-space bounded model and the Harmonic algorithm [9] when the number of active bins goes to infinity. The questions concerning t-space bounded d-dimensional packing ($d \geq 2$) have been studied in a number of papers. For large number of active bins, Epstein and van Stee [1] presented a $(\Pi_\infty)^d$-competitive space bounded algorithm, where $\Pi_\infty \approx 1.69103$ is the competitive ratio of the one-dimensional algorithm Harmonic. Algorithms for 2-dimensional bin packing with only one active bin were explored for the first time in [12], where the authors give 8.84-competitive algorithm for 2-imensional bin packing. An improved result of that case can be found in the paper [11], where a 5.155-competitive method is presented. The last article also contains an algorithm for packing squares with competitive ratio at most 4.5. In [7], a 4.84-competitive 1-space bounded 2-dimensional bin packing algorithm was presented. Grzegorek and Januszewski [3] presented a 3.5^d-competitive as well as a $12 \cdot 3^d$-competitive online d-dimensional hyperbox packing algorithm with one active bin. The d-dimensional case of 1-space bounded hypercube packing was discussed in [13], where a 2^{d+1}-competitive algorithm was described.

The case when there are two active bins was investigated in [4], where a 4.4138-competitive 2-space bounded 2-dimensional on-line packing algorithm was presented. Januszewski in [5] gave a 4-competitive 2-space bounded 2-dimensional algorithm and a 3.8-competitive 2-space bounded square packing method. The result was improved in [6], where two algorithms was described: a 3.8165-competitive 2-space bounded 2-dimensional on-line packing algorithm and a 2-space bounded method for packing squares with the competitive ratio 3.6. Shen and Zhao [10] extended the idea of 2-space bounded square packing described in [5] to 2-space bounded method of cube and hypercube packing. They first gave the 2-space bounded d-dimensional algorithms with the competitive ratio 5.43 for $d = 3$ and $32/21 \cdot 2^d$ for $d \geq 4$.

The aim of this paper is to improve both the upper bound (2^{3+1}) in the 3-dimensional case of 1-space bounded cube packing and the upper bounds of 2-space bounded methods of cube and hypercube packing presented in [10].

1.2 Our Results

The algorithm presented in Sect. 2 considers packing items (cubes of edges not greater than 1) into one active cube of edge 1. The main packing method is a bit like the classic computer game Tetris. The packing method which we describe is similar to the method presented by Grzegorek and Januszewski in [2]. The algorithm distinguishes types of items what determines a method for packing a specific item in a bin. Items that are considered big enough are packed from top to bottom. Different types of small items are packed from bottom upwards. The algorithm handles small items in a Tetris manner: to determine a place to pack an item a part of a bin is temporarily divided into congruent cuboids of appropriate size. Then an item is packed as low as possible inside a carefully chosen cuboid.

In Sect. 2 we give a 1-space bounded cube packing algorithm with the ratio 10.872.

In Sect. 3 we present a 2-space bounded cube and d-dimensional hypercube packing algorithm with the asymptotic competitive ratio $(3/2)^d + 2$ for $d \geq 3$. Our method of packing is natural and very simple and the proof of the upper bound theorem is easier than proofs presented in [10].

2 The One-Space-Algorithm

Let S be a sequence of cubes $Q_1, Q_2, \ldots$. Denote by a_i the edge length of Q_i.

- an item Q_i is *huge*, provided $a_i > 1/2$;
- an item Q_i is *big*, provided $1/4 < a_i \leq 1/2$;
- an item Q_i is *small*, provided $a_i \leq 1/4$; a small item Q_i is of *type* k provided $2^{-k-1} < a_i \leq 2^{-k}$.

Let $\mathcal{B}$ be the active bin. To shorten the notation, a cuboid whose edges have lengths $a \times a \times b$ will be called an (a, b)-cuboid.

2.1 Description of the One-Space-Algorithm

1. In packing items we distinguish coloured and white (not coloured) space. Items are placed only in the white space. Each newly opened bin is white.
2. We divide each freshly opened bin into $(1/2, 1)$-cuboids. These cuboids are named R_1, R_2, R_3, R_4 in an arbitrary order.
3. Huge items (edge $> 1/2$) are packed alone into a bin, i.e., if Q_i is huge, then we close the active bin and open a new bin to pack this item. After packing Q_i we close the bin and open a new active bin.

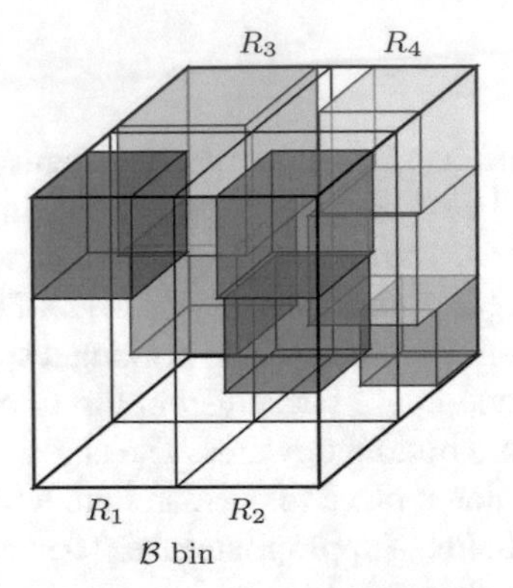

Fig. 1 Big items—the darker an item's colour, the later it arrived

4. If Q_i is big ($1/4 < \text{edge} \leq 1/2$) we find the highest indexed R_j such that Q_i can be packed into it. We pack Q_i into R_j along the edge of $\mathcal{B}$ as high as it is possible (see Figs. 1 and 3). If such a packing is not possible, we close the active bin, open a new active bin and pack Q_i into it.
   When a big item is packed, it colours the space covered by itself.
5. If Q_i is a small item of type k ($2^{-k-1} < \text{edge} \leq 2^{-k}$) (see Figs. 2 and 3) we find the lowest indexed R_j such that Q_i can be packed into it. Since j is fixed now, we will write R instead of R_j.
   We temporarily divide R into $(2^{-k}, 1)$-cuboids called $R(1), \ldots, R(4^{k-1})$. Denote by $t(n)$ the distance between the top of $R(n)$ and the top of the topmost item packed in $R(n)$ for $n = 1, \ldots, 4^{k-1}$ (see Fig. 4, right) and let η be an integer such that $t(\eta) = \max\{t(1), \ldots, t(4^{k-1})\}$.
   We pack Q_i into $R(\eta)$ as low as possible. The result of packing Q_i is the colouring of the $(2^{-k}, 1 - t(\eta) + a_i)$-cuboid contained in the bottom of $R(\eta)$ (see Fig. 4, right, where $\eta = 2$ before Q_{14} was packing).
   If such a packing is not possible, then we close the active bin and open a new active bin to pack Q_i.

2.2 Competitive Ratio

Let P_j for $j = 1, \ldots, 16$ be $(1/4, 1)$-cuboids with pairwise disjoint interiors. Each cuboid R_i for $i \in \{1, 2, 3, 4\}$ is divided into four cuboids $P_{4i-3}, \ldots, P_{4i}$ (see Fig. 5).

Lemma 1 *Assume that only small items were packed into $\mathcal{B}$. Assume that $j \in \{1, 2, \ldots, 16\}$. Denote by n the number of items packed into P_j and by t_n the distance between the bottom of $\mathcal{B}$ and the top of the topmost item packed into P_j. The total volume v_n of small items packed into P_j is greater than*

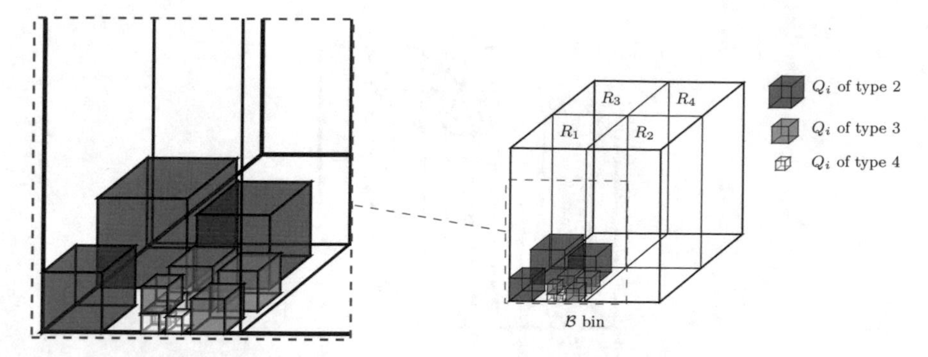

Fig. 2 Small items

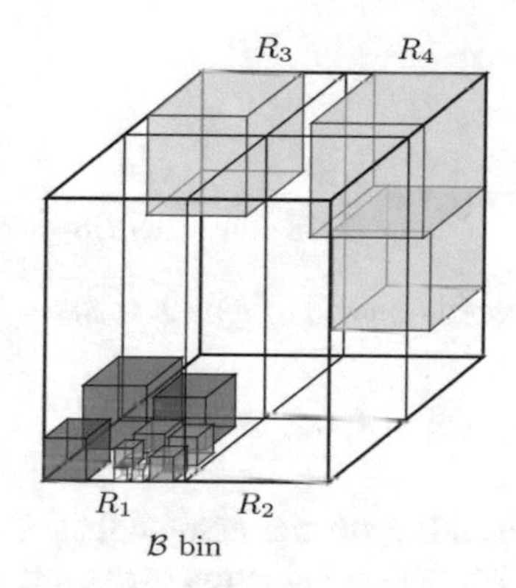

Fig. 3 One space-algorithm

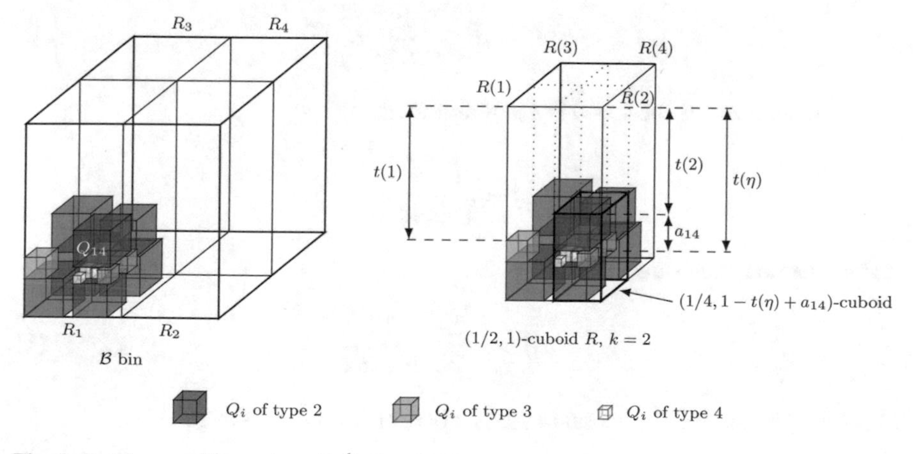

Fig. 4 Packing small items into $(2^{-k}, 1)$-cuboids

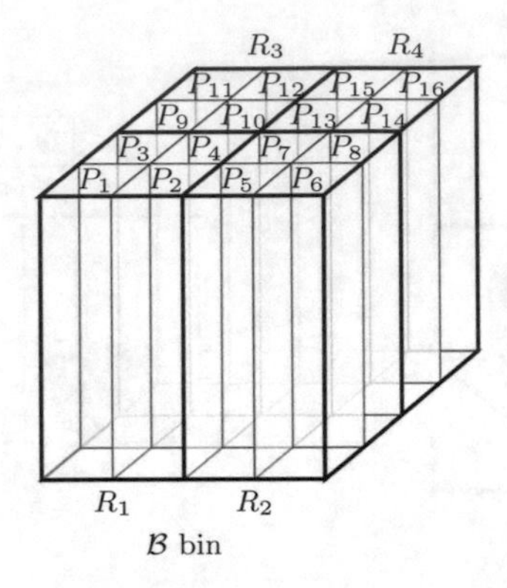

Fig. 5 (1/4, 1)-cuboids P_j

$$f(t_n) = \frac{19}{2048} \cdot t_n - \frac{13}{16,384}.$$

Moreover, if the topmost packed item is of type 2, *then*

$$v_n > f_+(t_n) = \frac{19}{2048} \cdot t_n.$$

Proof Without loss of generality we can assume that $P_j = [0, 1/4] \times [0, 1/4] \times [0, 1]$. We will prove the result using induction over the number n of packed items.

First assume that only one item Q_b was packed into P_j. Obviously, $t_1 = a_b$. Let

$$\varphi(a) = a^3 - \frac{19}{2048}a.$$

The function $\varphi(a)$ for $a > 0$ has a minimum at

$$a_0 = \sqrt{\frac{19}{6144}}.$$

A computation shows that

$$\varphi(a_0) > -\frac{1}{2} \cdot \frac{13}{16,384} \tag{1}$$

(this lower bound will be useful in the last part of the proof). We get

$$v_1 = a_b^3 > \frac{19}{2048} \cdot t_1 - \frac{1}{2} \cdot \frac{13}{8192} = f(t_1).$$

Moreover, if $1/8 < a_b \leq 1/4$, then $v_1 = a_b^3 > \frac{19}{2048} a_b = f_+(t_1)$

Now assume that the statement holds for at most n items packed into P_j (this is our inductive assumption). Let Q_u be the $(n+1)$st item packed into P_j and let t_{n+1} be the distance between the bottom of P_j and the top of the topmost item (from among $n+1$ items $Q_b, \ldots, Q_u$) packed into P_j.

If $a_u > 1/8$, then $t_{n+1} = t_n + a_u$. Using the inductive assumption,

$$v_{n+1} = v_n + a_u^3 > f(t_n) + a_u^3 = \frac{19}{2048} \cdot t_n - \frac{13}{16,384} + a_u^3.$$

Since

$$\varphi'(a) = 3a^2 - \frac{19}{2048} > 3 \cdot \frac{1}{64} - \frac{19}{2048} > 0$$

for $a > 1/8$, we get

$$\varphi(a) > \varphi\left(\frac{1}{8}\right) = \frac{13}{16,384}$$

for $a > 1/8$. Consequently,

$$\begin{aligned} v_{n+1} &> f(t_n) + a_u^3 = \frac{19}{2048}(t_n + a_u) + \varphi(a_u) - \frac{13}{16,384} \\ &\geq \frac{19}{2048}(t_n + a_u) = f_+(t_n + a_u) = f_+(t_{n+1}). \end{aligned}$$

Finally, consider the case when $a_u \leq 1/8$. First, we choose the topmost packed item Q^1 with edge greater than $1/8$ and denote by τ the distance between the bottom of P_j and the top of Q^1 (see Fig. 6, left). If there is no such item, then we take $\tau = 0$. The total volume of items packed up to τ, by the inductive assumption, is not smaller than $f_+(\tau)$. Above Q^1 we divide P_j into four $(1/8, 1-\tau)$-cuboids $P_j^1, P_j^2, P_j^3, P_j^4$. Denote by $Q_1^l, \ldots, Q_{v_l}^l$ the items from among $Q_b, \ldots, Q_{u-1}$ packed into P_j^l above Q^1 (if any) for each $l = 1, 2, 3, 4$. Moreover, denote by t_n^l the distance between the bottom of P_j and the top of the topmost item from among $Q_b, \ldots, Q_{u-1}$ packed into P_j^l and let $t_n^* = \min(t_n^1, t_n^2, t_n^3, t_n^4)$ (see Fig. 6, right). Clearly, $t_n^* \geq \tau$ and $t_n^* \leq t_n$.

If $t_n^* + a_u \leq t_n$, then $t_{n+1} = t_n$. Consequently,

$$v_{n+1} \geq v_n + a_u^3 = f(t_n) + a_u^3 = f(t_{n+1}) + a_u^3 > f(t_{n+1}).$$

If $t_n^* + a_u > t_n$, then $t_{n+1} = t_n^* + a_u$. Items $Q_1^l, \ldots, Q_v^l$ were packed into $(1/8, t_n^l - \tau)$-cuboid P_j^l. Let $h(P_j^l) = [0, 1/4] \times [0, 1/4] \times [0, 2t_n^l - 2\tau]$ be the image of P_j^l in a homothety h of ratio 2. By the inductive assumption, the total volume of cubes $h(Q_1^l), \ldots, h(Q_v^l)$ is not smaller than $\frac{19}{2048}(2t_n^l - 2\tau) - \frac{13}{16,384} = f(2t_n^l - 2\tau)$. Since the volume of each $h(Q_i^l)$ is 8 times greater than the volume of Q_i^l, it follows that the total volume of cubes $Q_1^l, \ldots, Q_v^l$ is not smaller than $\frac{1}{8}f(2t_n^l - 2\tau)$.

Consequently,

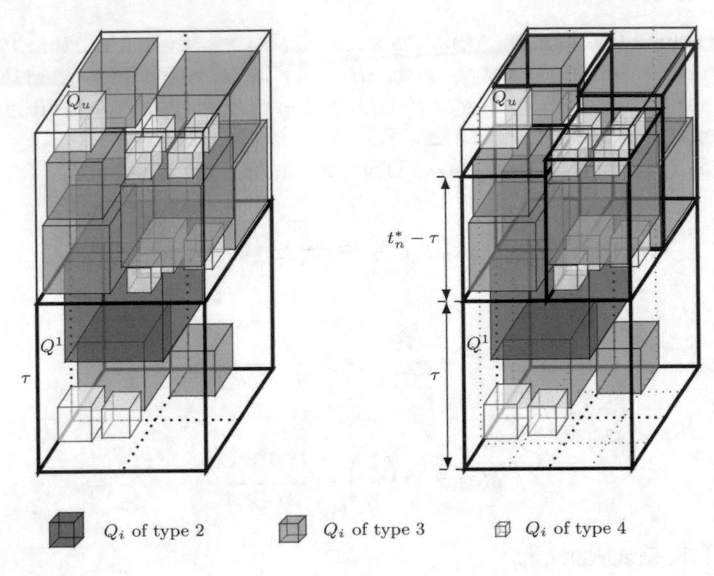

Fig. 6 The division

$$\begin{aligned} v_{n+1} &\geq f_+(\tau) + 4 \cdot \frac{1}{8} f\left(2t_n^* - 2\tau\right) + a_u^3 \\ &= a_u^3 + \frac{19}{2048} t_n^* - \frac{1}{2} \cdot \frac{13}{16,384}. \end{aligned}$$

By (1) we know that

$$\varphi(a_0) > -\frac{1}{2} \cdot \frac{13}{16,384}.$$

Consequently,

$$\begin{aligned} v_{n+1} &\geq \varphi(a_u) + \frac{19}{2048}(t_n^* + a_u) - \frac{1}{2} \cdot \frac{13}{16,384} \\ &> \frac{19}{2048}(t_n^* + a_u) - \frac{13}{16,384} = f(t_{n+1}). \end{aligned}$$

□

Lemma 2 *Define* $V_3 = 101/1024$. *Let S be a finite sequence of cubes and let* ν *be the number of bins used to pack items from S by the one-space-algorithm. Moreover, let m be the number of huge items in S. The total volume of items in S is greater than* $2^{-3} \cdot m + V_3(\nu - 2m - 1)$.

Proof Among ν bins used to pack items from S by the *one-space*-algorithm the first $\nu - 1$ bins will be called *full*. Let Q_z be the first item from S which cannot be packed into a full bin $\mathcal{B}$ by the *one-space*-algorithm. Clearly, Q_z is the first item packed into the next bin.

Denote by $v_{\mathcal{B}}$ the sum of volumes of items packed into $\mathcal{B}$.

If the incoming item Q_z is huge, then the average occupation ratio in both bins $\mathcal{B}_j$ and the next bin $\mathcal{B}_{j+1}$ into which Q_z was packed is greater than $1/2^4$. Obviously, there are $2m$ such bins.

It is possible that the last bin is almost empty.

To prove Lemma 2 it suffices to show that if Q_z is not huge and if no huge item was packed into $\mathcal{B}$, then $v_{\mathcal{B}} > V_3$ (the number of such bins equals $\nu - 2m - 1$).

Case 1: *Q_z is small and all items packed into $\mathcal{B}$ are small.*

Since $a_z \leq 1/4$, it follows that each P_i is packed up to height at least 3/4. By Lemma 1 we deduce that

$$v_{\mathcal{B}} > 4^2 f\left(\frac{3}{4}\right) = 16 \cdot \left(\frac{19}{2048} \cdot \frac{3}{4} - \frac{13}{4 \cdot 16,384}\right) = V_3.$$

Case 2: *Q_z is small and a big item was packed into $\mathcal{B}$.*

The volume of a big item Q_b with edge t is equal to $t^3 > t \cdot \left(\frac{1}{4}\right)^2$. In considerations presented in Case 1 we accept that the total volume of small items packed into R_j up to height t equals $4f(t)$. It is easy to see that

$$4f(t) < \frac{1}{16} \cdot t.$$

As a consequence, $v_{\mathcal{B}} > V_3$.

Case 3: *Q_z is a big item and all items packed into $\mathcal{B}$ are small*

Assume that there is $(2^{-2}, 1)$-cuboid $R_j(n)$ $(j \in \{1, 2, 3\},\ n \in \{1, 2, 3, 4\})$ such that the distance between its top and the top of the topmost item packed into it is greater than 1/8 and denote by R_+ first such cuboid. The total volume of items packed into R_+ is greater than $f(3/4)$. The total volume of items packed into each cuboid preceding R_+ is greater than $f(7/8)$. The total volume of items packed into each of remaining cuboids is greater than $\frac{3}{4} \cdot \frac{1}{8^2} > f(7/8)$ (in such a cuboid only items greater than $1/8$ were packed).

Denote by Q_n the topmost small item packed in R_4 (as in Fig. 7). Since $a_z \leq 1/2$ and Q_z cannot be packed in R_4, it follows that

$$v_{\mathcal{B}} > (16 - 5)f\left(\frac{7}{8}\right) + f\left(\frac{3}{4}\right) + 4f\left(\frac{1}{2} - a_n\right) + a_n^3.$$

Denote by $\gamma(a_n)$ the function on the right-hand side of this formula. This function for positive a has a minimum at $a_0 = \sqrt{\frac{19}{1536}}$.

A computation shows that $\gamma(a_0) > V_3$. Consequently, $v_{\mathcal{B}} > V_3$.

If there is no $(2^{-2}, 1)$-cuboid $R_j(n)$ $(j \in \{1, 2, 3\},\ n \in \{1, 2, 3, 4\})$ such that the distance between its top and the top of the topmost item packed into it is greater than $1/8$, then

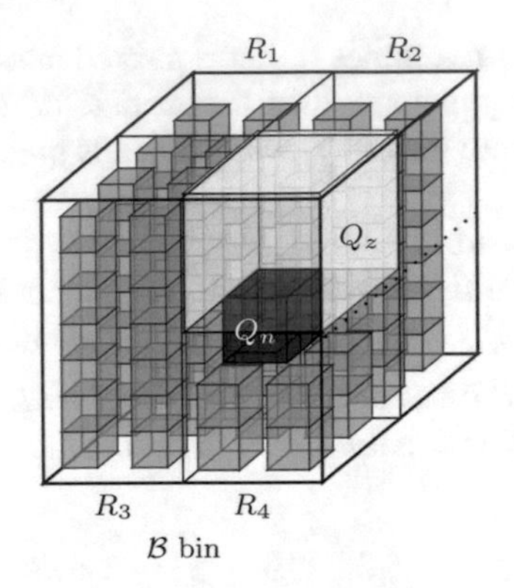

Fig. 7 Case 3

$$v_{\mathcal{B}} > (16-4)f\left(\frac{7}{8}\right) + 4f\left(\frac{1}{2} - a_n\right) + a_n^3.$$

Since $f(7/8) > f(3/4)$, we get $v_{\mathcal{B}} > \gamma(a_0) > V_3$.

Case 4: *Q_z is big and a big item was packed into $\mathcal{B}$*

Similarly as in Case 2 we get

$$4f(t) < t^3.$$

We deduce by Case 3 that $v_{\mathcal{B}} > V_3$.

Theorem 1 *The asymptotic competitive ratio for the one-space-algorithm is not greater than* $1098/101 \approx 10.8713$.

Proof Let S be a sequence of items of total volume v, let m denote the number of huge items in S and let μ be the number of bins used to pack items from S using the *one-space*-algorithm. Obviously, $OPT(S) \geq v$ as well as $OPT(S) \geq m$.

By Lemma 2 we get $v > \frac{1}{2^3} \cdot m + V_3 \cdot (\mu - 2m - 1)$, i.e.,

$$\mu < \frac{v}{V_3} + m\left(2 - \frac{1}{2^3 V_3}\right) + 1.$$

It is easy to check that $2 - \frac{1}{8V_3} > 0$.

If $m < v$, then

$$\frac{\mu}{OPT(S)} \leq \frac{\mu}{v} < \frac{\frac{v}{V_3} + v\left(2 - \frac{1}{2^3 V_3}\right) + 1}{v} = \frac{2^3 - 1}{2^3 V_3} + 2 + \frac{1}{v}.$$

If $v \leq m$, then

$$\frac{\mu}{OPT(S)} \leq \frac{\mu}{m} \leq \frac{\frac{m}{V_3} + m\left(2 - \frac{1}{2^3 V_3}\right) + 1}{m} = \frac{2^3 - 1}{2^3 V_3} + 2 + \frac{1}{m}.$$

Consequently, the asymptotic competitive ratio for the *one-space*-algorithm is not greater than

$$\frac{7}{8}\cdot\frac{1024}{101}+2=\frac{1098}{101}<10.872.$$

□

3 The *two-space*-algorithm

Let k be a non-negative integer and let $d \geq 3$. Denote by a_i the edge length of a d-dimensional hypercube C_i, for $i = 1, 2 \ldots$. Hypercubes are divided into types:

- a hypercube C_i is *big* provided $a_i > 1/2$;
- a hypercube C_i is *small* provided $a_i \leq 1/2$;
  - a small item C_i is of *type* $(2, k)$ provided $\frac{1}{3}\cdot 2^{-k} < a_i \leq 2^{-k-1}$;
  - a small item C_i is of *type* $(3, k)$ provided $2^{-k-2} < a_i \leq \frac{1}{3}\cdot 2^{-k}$.

Small items of *type* $(2, k)$ will be packed into a $\mathcal{B}_2$-bin while small items of *type* $(3, k)$ will be packed into a $\mathcal{B}_3$-bin.

Each $\mathcal{B}_2$-bin can be partitioned into 2^d congruent hypercubes called *2-subhypercubes*. Each $\mathcal{B}_3$-bin can be partitioned into 3^d congruent hypercubes called *3-subhypercubes*. Moreover, each m-subhypercube can be partitioned into 2^d congruent hypercubes, also called *m-subhypercubes* for $m \in \{2, 3\}$.

Given a hypercube C_i of type (m, k), denote by $K(C_i)$ the smallest m-subhypercube into which C_i can be packed.

Let $m \in \{2, 3\}$. An m-subhypercube of $\mathcal{B}_m$ is *i-empty* if its interior has an empty intersection with any packed hypercube $C_1, \ldots, C_{i-1}$. If $i = 1$, then all m-subhypercubes are empty.

3.1 Algorithm Small$(C_i, \mathcal{B}_m)$

Algorithm *small*$(C_i, \mathcal{B}_m)$ **for packing of small item C_i of type (m, k) into $\mathcal{B}_m$.**

1. If there is no i-empty m-subhypercube congruent to or larger than $K(C_i)$, then we stop the process of packing hypercubes into $\mathcal{B}_m$.
2. Otherwise, C_i is packed in $\mathcal{B}_m$ as follows:
   - if there is an i-empty m-subhypercube congruent to $K(C_i)$, then we pack C_i into it;
   - otherwise, we partition the smallest empty m-subhypercube that is larger than $K(C_i)$ into 2^d congruent m-subhypercubes $K_1, \ldots, K_{2^d}$ and we pack C_i into K_1 (see Fig. 8, left).

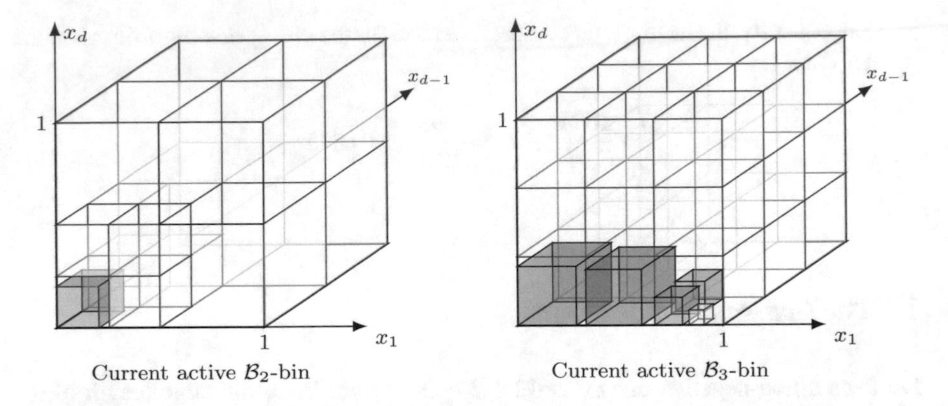

Fig. 8 Packing small items of type (m, k) into active $\mathcal{B}_m$-bin

Denote by $|A|$ the d-dimensional volume of A.

Lemma 3 *If C_i is packed in an m-subhypercube congruent to $K(C_i)$, then at least $\left(\frac{2}{3}\right)^d$ of this subhypercube is occupied.*

Proof If C_i is of type $(2, k)$, then

$$\frac{|C_i|}{|K(C_i)|} > \frac{(3^{-1} \cdot 2^{-k})^d}{(2^{-k-1})^d} = \left(\frac{2}{3}\right)^d.$$

If C_i is of type $(3, k)$, then

$$\frac{|C_i|}{|K(C_i)|} > \frac{(2^{-k-2})^d}{(3^{-1} \cdot 2^{-k})^d} = \left(\frac{3}{4}\right)^d > \left(\frac{2}{3}\right)^d.$$

□

Lemma 4 *Let $m \in \{2, 3\}$. If the algorithm small cannot pack a small hypercube C_i into $\mathcal{B}_m$, then the sum of volumes of hypercubes packed into $\mathcal{B}_m$ is greater than $\rho = (2^d - 1)/3^d$.*

Proof Let C_i be a small hypercube of *type* (m, k). All i-empty m-subhypercubes in $\mathcal{B}_m$ are smaller than $K(C_i)$. Furthermore, for each $j = 1, 2, \ldots$ there are at most $2^d - 1$ empty m-subhypercubes congruent to $2^{-j}K(C_i)$. The total volume of i-empty m-subhypercubes in $\mathcal{B}_m$ is smaller than

$$(2^d - 1) \cdot \left[\left(\frac{1}{2}\right)^d + \left(\frac{1}{4}\right)^d + \ldots\right] \cdot |K(C_i)| = |K(C_i)|.$$

This implies that the total volume of no-empty m-subhypercubes is greater than $1 - |K(C_i)|$. By Lemma 3 we know that the total volume of hypercubes $C_1, \ldots, C_{i-1}$

is greater than $\left(\frac{2}{3}\right)^d (1 - |K(C_i)|)$. Since $|K(C_i)| \leq \frac{1}{2^d}$, we get

$$\sum_{n=1}^{i} |C_n| > \left(\frac{2}{3}\right)^d \left(1 - \frac{1}{2^d}\right) = \frac{2^d - 1}{3^d} = \rho.$$

□

3.2 Algorithm Subhypercube

At any time of the packing process two bins are active: $\mathcal{B}_3$ for packing items of *type* $(3, k)$ and $\mathcal{B}_2$ for packing items of *type* $(2, k)$ as well as for packing big items.

Let $C_1, C_2, \ldots$ be a sequence of d-dimensional hypercubes.

Algorithm *subhypercube*.

1. If C_i is big, then we close the active $\mathcal{B}_2$ bin, open a new bin, pack C_i and close the bin (see Fig. 9). Then we open a new active $\mathcal{B}_2$ bin.
2. If C_i is a small item of *type* $(2, k)$, then we pack it in the active $\mathcal{B}_2$ bin by $small(C_i, \mathcal{B}_m)$. If such a packing is not possible, we close the active $\mathcal{B}_2$ bin and open a new active $\mathcal{B}_2$ bin to pack C_i.
3. If C_i is a small item of *type* $(3, k)$, then we pack it in the active $\mathcal{B}_3$ bin by $small(C_i, \mathcal{B}_m)$. If such a packing is not possible, we close the active $\mathcal{B}_3$ bin and open a new active $\mathcal{B}_3$ bin to pack C_i (see Fig. 10, where $k = 0$).

Theorem 2 *The asymptotic competitive ratio of the subhypercube-algorithm is not greater than* $\left(\frac{3}{2}\right)^d + 2$.

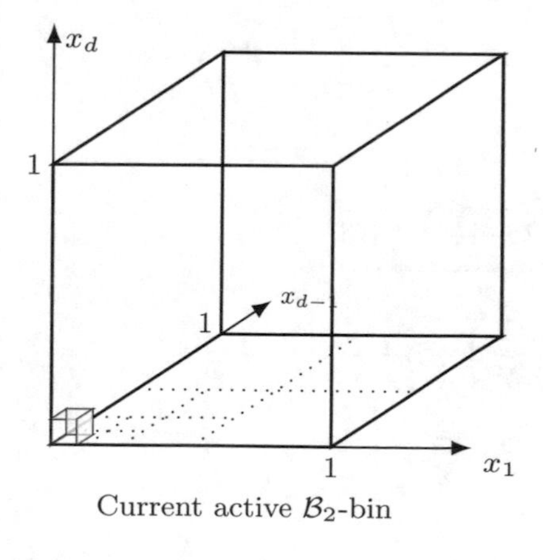

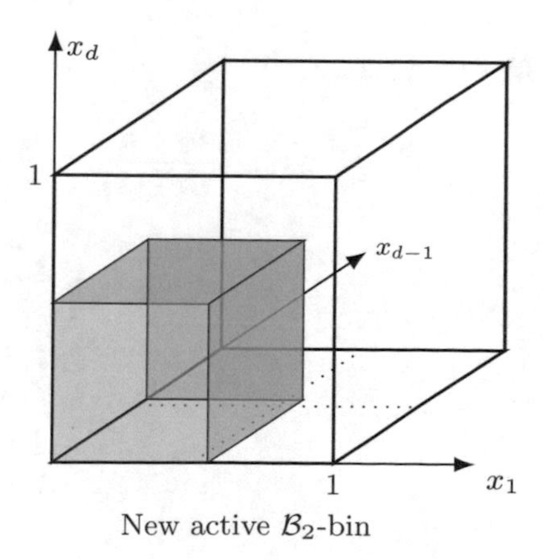

Fig. 9 Packing a big hypercube into $\mathcal{B}_2$-bin

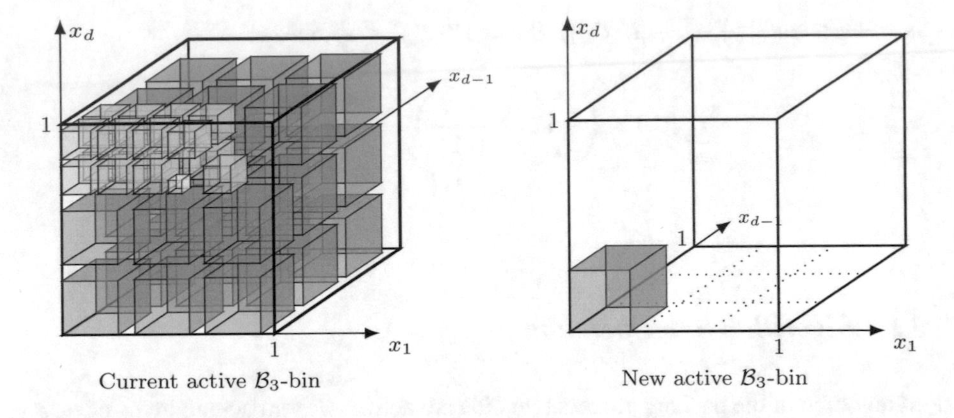

Fig. 10 Packing a small item of type (3, 0) into $\mathcal{B}_3$-bin

Proof Let S be a sequence of items of total volume v, let n denote the number of big items in S and let n_j be the number of $\mathcal{B}_j$ bins ($j \in \{2, 3\}$) used to pack items from S with *subhypercube*-algorithm. If the incoming item C_i is big, then the average occupation ratio in both current active $\mathcal{B}_2$-bin and the next $\mathcal{B}_2$-bin into which C_i was packed is greater than $1/2^{d+1}$ (see Fig. 9). Obviously, there are $2n$ such bins.

It is possible that the last $\mathcal{B}_2$ and $\mathcal{B}_3$ bins are almost empty.

Obviously, $OPT(S) \geq v$ as well as $OPT(S) \geq n$. By Lemma 4 we get

$$v > \frac{1}{2^d} \cdot n + \rho \cdot (n_2 + n_3 - 2n - 2),$$

i.e.,

$$n_2 + n_3 < \frac{1}{\rho} v + \left(2 - \frac{1}{2^d \rho}\right) n + 2.$$

If $n < v$, then

$$\begin{aligned}
\frac{n_2 + n_3}{OPT(S)} &< \frac{n_2 + n_3}{v} \\
&< \frac{\frac{1}{\rho} v + \left(2 - \frac{1}{2^d \rho}\right) v + 2}{v} \\
&= \frac{1}{\rho}\left(1 - \frac{1}{2^d}\right) + 2 + \frac{2}{v} \\
&= \left(\frac{3}{2}\right)^d + 2 + \frac{2}{v}.
\end{aligned}$$

If $v \leq n$, then

$$\begin{aligned}
\frac{n_2 + n_3}{OPT(S)} &< \frac{n_2 + n_3}{n} \\
&\leq \frac{\frac{1}{\rho}n + \left(2 - \frac{1}{2^d \rho}\right)n + 2}{n} \\
&= \frac{1}{\rho}\left(1 - \frac{1}{2^d}\right) + 2 + \frac{2}{n} \\
&= \left(\frac{3}{2}\right)^d + 2 + \frac{2}{n}.
\end{aligned}$$

Consequently, the asymptotic competitive ratio for the subhypercube algorithm is not greater than $\left(\frac{3}{2}\right)^d + 2$.

□

References

1. Epstein, L., van Stee, R.: Optimal online algorithms for multidimensional packing problems. SIAM J. Comput. **35**(2), 431–448 (2005)
2. Grzegorek, P., Januszewski, J.: A note on one-space bounded square packing. Inf. Process. Lett. **115**(11), 872–876 (2015)
3. Grzegorek, P., Januszewski, J.: Drawer algorithms for 1-space bounded multidimensional hyperbox packing. J. Comb. Optim. **37**(3), 1011–1044 (2019)
4. Hu, S., Wang, X.: An algorithm for online two-dimensional bin packing with two open bins. J. Inf. Comput. Sci. **10**(17), 5449–5456 (2013)
5. Januszewski, J.: On-line algorithms for 2-space bounded 2-dimensional bin packing. Inf. Process. Lett. **112**(19), 719–722 (2012)
6. Januszewski, J., Zielonka, Ł.: Improved online algorithms for 2-space bounded 2-dimensional bin packing. Int. J. Found. Comput. Sci. **27**(4), 407–429 (2016)
7. Januszewski, J., Zielonka, Ł.: Online packing of rectangular items into square bins. In: Solis-Oba, R., Fleischer, R. (eds.) Approximation and Online Algorithms. WAOA 2017, volume 10787 of Lecture Notes in Computer Science, pp. 147–163. Springer, Cham (2018)
8. Johnson, D.S.: Fast algorithms for bin packing. J. Comput. Syst. Sci. **8**(3), 272–314 (1974)
9. Lee, C.-C., Lee, D.-T.: A simple on-line bin-packing algorithm. J. ACM **32**(3), 562–572 (1985)
10. Shen, H., Zhao, X.: On-line algorithms for 2-space bounded cube and hypercube packing. Tsinghua Sci. Technol. **20**(3), 255–263 (2015)
11. Zhang, Y., Chen, J., Chin, F.Y.L., Han, X., Ting, H.-F., Tsin, Y.H.: Improved online algorithms for 1-space bounded 2-dimensional bin packing. In: Cheong, O., Chwa, K.-Y., Park, K. (eds.) Algorithms and Computation, pp. 242–253. Springer, Berlin (2010)
12. Zhang, Y., Chin, F.Y.L., Ting, H.-F.: One-space bounded algorithms for two-dimensional bin packing. Int. J. Found. Comput. Sci. **21**(6), 875–891 (2010)
13. Zhang, Y., Chin, F.Y.L., Ting, H.-F., Han, X.: Online algorithms for 1-space bounded multi dimensional bin packing and hypercube packing. J. Comb. Optim. **26**(2), 223–236 (2013)

Author Index

A
Arakelyan, Alexander, 1

C
Cabodi, Gianpiero, 71
Camurati, Paolo, 71
Cardil, Adrián, 33

D
Dobrinkova, Nina, 1, 33
Dörpinghaus, Jens, 49
Düing, Carsten, 49

F
Fadda, Edoardo, 71
Fidanova, Stefka, 89

G
Ganzha, Maria, 89

H
Hirata, Kouichi, 111

I
Ivanov, Vladimir, 173

K
Katsaros, Evangelos, 1

M
Manerba, Daniele, 71

P
Pavlova, Kristina, 173

R
Rapoport, Michael, 149
Reynolds, Sean, 1
Roeva, Olympia, 89, 135

T
Tadei, Roberto, 71
Tamir, Tami, 149

U
Ukita, Yoshiyuki, 111

W
Weil, Vera, 49

Y
Yoshino, Takuya, 111

Z
Zielonka, Łukasz, 183
Zoteva, Dafina, 135

S. Fidanova (ed.), *Recent Advances in Computational Optimization*,
Studies in Computational Intelligence 920,
https://doi.org/10.1007/978-3-030-58884-7

Printed in the United States
by Baker & Taylor Publisher Services